Who Built The Moon?

Alan Butler
Christopher Knight

EasyRead Large

Copyright Page from the Original Book

TABLE OF CONTENTS

Christopher Knight has worked in advertising and marketing for over thirty years, specialising in consumer psychology and market research. His writing career began almost by accident after he had invested seven years conducting research into the origins of Freemasonic rituals and he has written four books to date, co-authored with Robert Lomas. His first book, *The Hiram Key,* was published in 1996 and it immediately went into the UK top ten bestseller list and remained in the chart for eight consecutive weeks. It has since been translated into thirty-seven languages and sold over a million copies worldwide, becoming a bestseller in several countries. He now divides his time between marketing consultancy and historical research for writing books.

Alan Butler qualified as an engineer, but was always fascinated by history, and made himself into something of an expert in astrology and astronomy. Since 1990, he has been researching ancient cultures, pagan beliefs and comparative religion and has published four successful books on such topics as the Knights Templar and the Grail legend. He is also a published playwright and a very successful radio dramatist.

By the same authors

Previous books by Christopher Knight
(co-authored with Robert Lomas)
The Hiram Key
The Second Messiah
Uriel's Machine
The Book of Hiram

Previous books by Alan Butler
The Bronze Age Computer Disc
The Warriors and the Bankers
The Templar Continuum
The Goddess, the Grail and the Lodge

By Christopher Knight and Alan Butler
Civilization One

The publisher would like to thank the following people, museums, and photographic libraries for permission to reproduce their material. Every care has been taken to trace copyright holders. However, if we have omitted anyone we apologize and will, if informed, make corrections to any future edition.

Figure 4 Drawing of Calendar Stone – Courtesy of Philip Stooke, University of Western Ontario

List of plates

Dedication

For my Mother, my brother Peter
and in loving memory of my Father. CK.

For my good friends Henry and Michelle. AB.

Acknowledgments

Kate Butler, for her usual invaluable assistance with proofs and index.

Penny Stopa and the editorial team.

Fiona-Spencer Thomas, who was wonderful as always.

Hilary Newbigen, for her customary apposite comments and advice.

Michael Mann, who continues to encourage, advise, and assist.

Introduction

For most people the suggestion that the Moon could be artificial is about as sensible as saying that it is made of green cheese. This is a perfectly reasonable response based on everything that we know about the world we live in, where there are just two kinds of objects: those that are here because the random forces of the Universe – that we call 'nature' – caused them to exist; or because they were manufactured by human hand.

However super-rational our scientific community considers itself to be, there are still huge numbers of people who believe things that are not proven by empiric means. In a recent poll it was found that no less than ninety-two per cent of Americans say they believe in God[1] – and other surveys indicate that many millions of people are equally convinced that aliens have visited our planet.

God may well exist, and so too might aliens for all we know, but this book will only concern itself with hard, scientific facts. And, unlike so many of those trapped in the politically correct world of academia, our published findings will not be constrained by the demands of current convention. The information we put forward here is clear, testable and, we believe, irrefutable.

Despite the fact that the Moon is almost certainly 4.6 billion years old, we will demonstrate beyond all reasonable doubt that Earth's Moon cannot be a natural object. And then we shall explain in detail how the agency that manufactured the Moon left a series of detailed messages of what had been done and for whom it had been undertaken.

So, here is our challenge. Put aside your natural incredulity and read this book with an open mind, check out the evidence then ask yourself 'Who built the Moon?'

We have cited three possibilities but maybe you can think of more. However, the last of our suggestions appears to us to be increasingly likely. It is a worrying, staggering, exciting and completely awesome concept. And, if there is even an outside chance that this could be the answer, the world has a major new challenge ahead of it.

CHAPTER ONE

The Dawn of Awareness

The entire population appeared to have simultaneously decided to evacuate every building, and the streets and car parks were quickly filling with people standing almost shoulder-to-shoulder. Traffic began to grind to a halt as drivers leaned out of their windows, and even the birds abandoned the sky to assemble in rows along guttering and telephone wires, chattering like some misplaced dawn chorus.

The large grey clouds obligingly parted to reveal a muted latemorning Sun that had a small bite out of its right-hand edge. As the dark spot grew, the birds fell silent and a sea of expectant faces became transfixed upwards. Three welders from a nearby garage became instantly popular as they passed around their dark-lensed masks, allowing the smiling onlookers a direct view of the diminishing solar disc.

Then it happened; the moment of totality arrived. The Sun disappeared for several seconds, allowing the darkness of night to wholly consume the day. Then slowly a bright sparkle materialized that soon looked like a diamond set on the band of some heavenly ring.

The last total solar eclipse of the twentieth century had just occurred on the morning of August 11th 1999. It had begun when the Moon passed between the Earth and Sun, throwing an umbral shadow, forty-nine kilometres wide, on the North Atlantic just south of Nova Scotia. The inky black circle then swept across the ocean surface until it passed over the Isles of Scilly, off the south-west coast of England, some forty minutes later. Here the path width had expanded to 103 kilometres and was now covering the ground at a speed approaching 1,000 metres per second. The circular shadow then curved its way over Europe and on to the Middle East before crossing India and finally disappearing over the Bay of Bengal.

Such events do not happen often in the lifetime of an individual but once seen, a total solar eclipse is never forgotten. Solar eclipses occur around two to five times per year but the area on the ground covered by the totality is very small, so in any given location on Earth a total eclipse will only happen once every 360 years.

One can only imagine how primitive peoples may have feared for their lives as the Sun was apparently extinguished before their eyes. No doubt the astronomer priests of ancient time held sway over their people by having the apparently magical power of predicting such terrifying events.

But even today the magic and mystery of the eclipse is very real.

It is a very strange quirk of fate indeed that the disc of the Moon should seem, from an earthly perspective, to be exactly the same size as the Sun. Whilst we casually take it for granted that the two main bodies seen in Earth's skies look the same size, it is actually something of a miracle. Most people are fully aware that the Moon is tiny compared to the Sun but that it is much closer to us causing them to appear as equal discs. To be precise, the Moon is 400 times smaller than the star at the centre of our solar system, yet it is also just 1/400th of the distance between the Earth and the Sun.

Whilst the surprisingly neat number of 400 for relative size and distance is apparently an amusing coincidence of the decimal counting system, the odds against this optical illusion happening at all are huge. Experts are deeply puzzled by the phenomenon. Isaac Asimov, the respected scientist and science-fiction guru, described this perfect visual alignment as being 'the most unlikely coincidence imaginable'.

This perfect fit of the lunar and solar discs is a very human perspective because it only works from the viewpoint of someone standing on the Earth's surface. But the magic of the Moon's movements above our heads goes to even more astonishing levels. By some absolutely incomprehensible quirk of nature, the Moon also manages to precisely imitate the perceived annual movements of the Sun each month.

So, when the Sun is at its lowest and weakest in midwinter, the full Moon is at its highest and brightest, and in midsummer, when the Sun is at its highest and brightest, the Moon is at its weakest.

If you want to understand how extraordinary this doppelgänger effect is, stand on a hilltop or an open plain and film the Sun at midwinter sunset (its most southerly point on the horizon), at the spring equinox, again at midsummer and again at the autumn equinox. Then on those same dates film the Moon setting and you will see that they both go down at the same point on the horizon at the equinoxes (March 21st] and September 21st) but the Moon will have the opposite setting point to the Sun at solstices in December and June.[2]

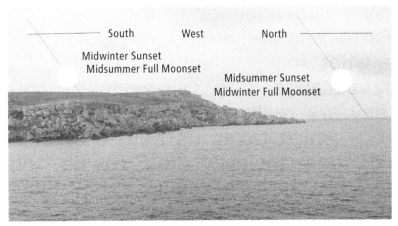

Figure 1. This drawing shows the peculiar relationship of the Sun and Moon throughout the year as seen from Earth. At midsummer in the northern hemisphere the Sun sets north of west, whereas the full Moon sets south of west. At midwinter the situation is reversed, with the Sun setting south of west and the Moon setting north of west.

Figure 2. At the time of the spring and autumn equinox sunset happens at a due west position, whilst the full Moon also sets in this part of the sky.

It would be easy to dismiss these Sun-mimicking performances by saying that it is simply a consequence of the Moon's distance from Earth and its orbital characteristics. And that is what most scientifically trained people will say because it is self-evidently true. But what they are really saying is 'It is so because it is so' – which takes us nowhere. Of course, it could, and logically has to be, one big coincidence. What else could it be? Even most of the ninety-two per cent of the American population who state that they believe in God would probably assume coincidence and only a minority might claim that it is the grand plan of the Almighty.

The Moon's dance around the Earth that produces these startling performances is extremely complex and it is a consequence of the relative movements of the Earth and the Sun as well of the Moon itself.

The path of the Moon's orbit is inclined at 5°9′ to the line of Earth's path around the Sun, known as the plane of the ecliptic. The Earth is also tilted at an angle of just over 23°27′, although this is slowly decreasing so that in several million years it will reach 22°54′, after which it will again increase.

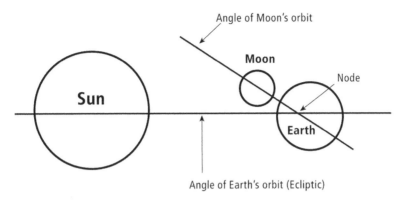

Figure 3

It follows that solar eclipses can only occur when the Moon passes through the plane of the ecliptic and the Sun's light is blocked by the Moon. These points of intersection happen twice for each lunar orbit and are known as 'nodes'. These nodes appear to move slowly around the background stars giving the impression of moving backwards through the calendar occurring 19.618 days earlier per year. The cycle completes every 18.618 years, which amounts to a surprisingly neat 6,800 days.

Closely allied to this node cycle is the so-called 'Saros cycle', which governs the periodicity and recurrence of eclipses, where each eclipse sequence has a duration of approximately 6,585.32 days (eighteen years, eleven days, seven hours, forty minutes and forty-eight seconds). The people of Ancient Mesopotamia knew of this astronomical principle and it is quite possible that earlier observers, long before written records began, were also aware of it.

However, one has to wait for three Saros cycles in order for a solar eclipse to repeat at the same spot on Earth because successive eclipses in the Saros cycle happen one-third of the way around the world from each other. You would therefore have to wait over fifty-four years to see an eclipse return to the same geographic area. There are twelve different Grand Saros eclipse series at the present time.

Human knowledge about the movements of the Moon is far older than most people might imagine. More than 25,000 years ago an early astronomer created a lunar calendar that is still intact. The bone he engraved was excavated nearly a hundred years ago at Abri Blanchard, not far from Lascaux in France. Experts agree that the markings accurately correspond with a two-month lunar calendar. Around 250 generations later another astronomer also recorded this already ancient knowledge, using various natural minerals daubed onto a cave wall to leave the image of an empty rectangle followed by a series of fourteen sooty dots. It was realized that these marks might also be a lunar calendar. The fourteen dots, it was argued, represented the face of the Moon from full to new, after which the empty rectangle would symbolize the disappearance of the Moon's face on the fifteenth day.

If anyone doubted that the marks on the cave walls at Lascaux really was a lunar calendar, or even continued to believe counting was something that did not appear until the arrival of the written word some 5,000 years ago, another picture close by might cause them to think again. On this part of the cave wall there were twenty-nine dots, snaking around the bottom of a beautifully executed painting of a wild horse. Twenty-nine days is the period from new Moon through full Moon to new Moon again. And yet another artefact known as the Isturitz Baton, displays an even more advanced four-month and five-month lunar calendar.

It is humbling to realize that these records were created more than ten thousand years before the Ice Age ended and the woolly mammoth disappeared.

These kinds of lunar observations are not restricted to southern France. The Ishango Bone, which was found in the Congo, Africa, also carries markings that seem to represent a lunar calendar. What is more, it is of an almost identical age to the Isturitz Baton, though it originated many hundreds of kilometres to the south and on a different continent.

The existence of lunar calendars from such an early date is of great importance to our understanding of our own development. They demonstrate a clear awareness of the passing of time and the cycles of the natural world. The discovery of an archaeological artefact is a matter of chance and is dependent on the number of objects of any particular sort that once existed. The fact that so many of these bones, antlers and paintings have been discovered is a good indication that they were not unique and that Moon knowledge was important to the Palaeolithic people of Europe and Africa, though this does give us cause to wonder why such an early lunar fascination developed.

A recent discovery has shown why such intricate observations 'suddenly' became possible for our distant forebears around 32,000 years ago. In July 2004, Rachel Caspari of the University of Michigan and Sang-Hee Lee of the University of California published a paper in the *Proceedings of the National Academy of Sciences,* concerning comparisons of 768 different human fossils from a huge span of human development. They then divided the fossils into two groups – adults of reproductive age, which they settled on as fifteen years, and adults that lived to be twice as old, based on tooth wear.

In primitive societies, people were often grandparents by the age of thirty, if they were lucky enough to live that long.

Dr Caspari said, 'We found this proportion of older to young adults in the fossil record increased over time and in the Upper Paleolithic that proportion just skyrocketed.'

By calculating the ratio of old to young individuals in the samples, the researchers found that their numbers soared up to fivefold in the Upper Palaeolithic group, a leap that was so surprising that the team at first questioned its own results.

This dramatic leap in average lifespan allowed individuals to grow older and wiser and afforded each of these new elders time to pass on their knowledge to the next generation of adults. The wear on the teeth suggests that this leap in longevity must have given rise to a true form of education that could build up a body of 'species intelligence' where the entire social group knows far more than any one individual. This would allow for the first specialization in which talented men and women were fed and protected by the group to allow them to add value to their early society.

This sudden transition from a society of children to one of 'greybeards' must have been a watershed that laid the foundations for what would eventually become true civilization. The period of history, known as the Upper Palaeolithic Period, marks a time when modern man was becoming established in Europe and there was an expansion of population, creating social pressures that led to the growth of trade networks, increased mobility, and more complex systems of co-operation and competition.

We could now understand why observational astronomy became the first real science for humankind. All science is based upon observation of patterns that stand out from the 'noise' of simple random chance and then, through understanding, we can make predictions of future events and outcomes. In this way the tides, the seasons and the movements of the heavens could be seen as being parts of a single engine driving the variations in the immediate environment of the early thinkers.

These early observational scientists would also note where patterns from completely different events appeared to be related. Why should high tide happen twice a day and rise higher when the Moon was full or when there was no Moon at all? Did the Moon have some kind of control over something as massive as the oceans? Even stranger, why did women of childbearing age lose blood once for every complete cycle of the Moon?

We can be sure that this particular fact was not lost upon these people.

Back in 1911 a French physician by the name of J G Lalanne was examining caves in Laussel, in the Dordogne, when he chanced upon something that turns out to be very illuminating in terms of the Palaeolithic mindset. Carved into the wall of a limestone rock shelter, he found a 33cm female figure. The artistry involved from so early a period is quite remarkable, the more so given that it was executed with flint tools. The naked and full-bellied woman has her left hand on her abdomen and in her right hand is holding a bison horn, in the shape of a crescent moon. Upon the bison horn there are thirteen incised lines. The Venus of Laussel, as she is called, is at least 20,000 years old.

This carving is one of many that strongly suggest there was a very early recognition that human fertility seemed to be tied to the phases and period of the Moon. Human female reproduction is dependent on the menstrual cycle which has an average of twenty-eight days, and approximately halfway through the cycle a mature cell is released from a woman's ovaries and becomes available for fertilization. If sexual intercourse does not take place and the egg is not fertilized, it disintegrates after a couple of days. At the end of the cycle, if no conception has taken place, menstruation begins and the cycle commences once again.

A series of intriguing studies by Professor LeRoy McDermott of the Missouri State University has suggested that these early 'Venus' images of the female figure were self-portraits. His analysis showed that the figurines were made from the point of view of 'self' rather than 'other' and they could only represent a women's view of her own body both emotionally and physically as she looks downwards. Using photographic simulations of what a modern female sees of herself, McDermott demonstrates that the anatomical omissions and proportional distortions found in various Venus figurines occur naturally in autogenous, or self-generated, information. The size, shape, and articulation of the objects appear to be determined by their relationship to the eyes and the relative effects of foreshortening, distance, and occlusion rather than by any symbolic distortion. As self-portraits of women at different stages of life, McDermott believes these earliest representations of the human form embodied obstetrical and gynaecological information and probably signified an advance in women's self-conscious control over the material conditions of their reproductive lives.

The lunar month symbolism in the Venus of Laussel strongly suggests that women 20,000 years ago knew the length of their menstrual cycles and already equated them with the phases of the Moon. The thirteen lines on the crescent-shaped bison horn could easily relate to the thirteen menstrual cycles an average woman could expect in each year. At the same time, it is not at all uncommon for a human female to menstruate on the same Moon phase each month because twenty-eight days is merely an average, whilst the period between one full Moon and the next is 29.53 days.

The historical connection between human fertility and the Moon even extends to the word 'menstrual'. It derives from the Latin *mensis,* meaning month, whilst the word 'month' is very ancient and refers to the period of four weeks as being one 'moonth'.

The connection between human fertility and the cycles of the Moon is considered to be 'apparent rather than actual', but it isn't in the least surprising that the possibility of a relationship was noticed by our ancient ancestors. The clincher probably came when someone realized that the average gestation period of a human female, from conception to birth, is around 266 days – or nine full lunar synodic cycles.

In a social and a religious sense, fertility undoubtedly played a crucial part in the lives of people at the time the Venus of Laussel was carved. It is more or less universally accepted that female deities were important to human culture for thousands of years of prehistory. Statues of pregnant women with exaggerated genitals and breasts are common from the Palaeolithic to the Neolithic periods and there are strong indications of the existence of a fertility-based deity who has come down to us as 'The Great Goddess'. The Venus of Laussel could quite easily be a representation of this deity, complete with a representation of the heavenly body with which she was equated – the Moon.

About 6,000 years ago there was an outbreak of building in stone across the western parts of Europe, particularly in the British Isles, that tells us a great deal about the Neolithic people's fascination with the Moon.

Dr Philip Stooke, of the University of Western Ontario, Canada had always been puzzled as to why there were no maps or drawings of the Moon older than the one drawn by Leonardo da Vinci five hundred years ago. He decided to look at ancient manuscripts and the records of excavations of the Neolithic sites on the British Isles. Amongst other sites, he looked at the truly amazing prehistoric structures known as Newgrange and Knowth in County Meath, Ireland. And it was at the recently excavated Knowth that he found a 5,200-year-old carving made up of a set of lines and dots. Dr Stooke realized that this was not simply a Stone-Age doodle but a drawing of the face of the Moon. He said:

'I was amazed when I saw it. Place the markings over a picture of the full Moon and you will see that they line up. It is without doubt a map of the Moon, the most ancient one ever found. It's all there in the carving. You can see the overall pattern of the lunar features, from features such as Mare Humorun through to Mare Crisium. The people who carved this Moon map were the first scientists – they knew a great deal about the motion of the Moon. They were not primitive at all.'

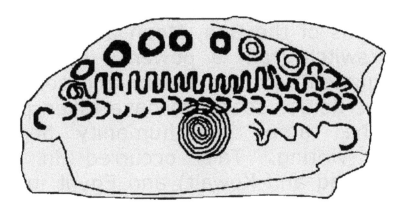

Figure 4

These people were not merely Moon watchers. Chris, along with Robert Lomas, had already published his analysis of the astronomical function of nearby Newgrange, which was carefully designed and engineered to allow the light of Venus to penetrate deep into the domed structure once every eighth winter solstice.[3] This focused beam of light gave these early scientists a very precise tracking of Venus, which allowed them to maintain a calendar that would be accurate to a matter of seconds over each eight-year cycle. There was no doubt that these builders were far from primitive, as archaeological convention once suggested.

Investigations at Knowth had already shown that at certain times moonlight shines down the eastern passage of the structure. Dr Stooke has now pointed out that these narrow moonbeams would also fall right onto the Neolithic lunar map. He concluded, 'It was obviously built by men who had a sophisticated understanding of the motions of the Sun, Moon and stars.'

The switch from a powerful female deity, often equated with the Moon, and solar-based masculine deities seems to have taken place at about the same time humanity began to discover writing. This occurred in Sumer (modern Iraq and Kuwait) and Egypt just after structures like Newgrange and Knowth had been constructed.

One researcher, Dr Leonard Shlain, Chief of Labroscopic Surgery at the California-Pacific Medical Center, has suggested this connection in his controversial but immensely popular book, *The Alphabet versus the Goddess.*[4] Here Shlain outlines his view that the evolution of writing specifically involved the use of the practical left hemisphere of the brain, as a direct contrast to the many thousands of years during which the more intuitive, inspirational right hemisphere had predominated. He maintains that this explains the virtual abandonment of a generally peaceful feminine-centred society across much of Europe, the Middle East and Asia. This transition was staggered but it began around 3,000BC, when a more aggressive, patriarchal social structure emerged with masculine deities predominating.

This thesis sounds very reasonable and, if true, we could expect to find this legacy of the Moon-associated goddesses still present at the dawn of writing, when myths and stories were first being catalogued. And this is indeed the case. In Sumer we find Nana, a very early Moon goddess, whilst in nearby Egypt, where writing came just a little later, there is an even better example in terms of Isis, who rose to become one of the most important and revered goddesses across the whole known world for several thousand years. Isis originated as a Moon goddess, and the fact is borne out by one specific part of her story. Isis had to rebuild the body of her husband, Osiris, after he had been brutally murdered and his body chopped into pieces. She travelled all over the world to find the dismembered parts of her husband of which there were fourteen in total. The story is analogous to the gradual increase in size of the Moon across fourteen days from new to full.

Referring to the Egyptians, Plutarch, the Greek essayist, writing around 60AD said:

> 'Egyptian priests called the Moon "the Mother of the Universe", because the Moon, having the light which makes moist and pregnant, is promotive of the generation of living beings.'

Although to some early cultures the Moon was associated with a masculine deity, such as the Babylonian Sin for example, in by far the majority of cases the Moon was considered to be female and carried strong aspects of fertility. This goddess had many names across the world. To the Greeks she was Artemis and the Romans called her Diana and Selene. Her Finnish name was Kuu and to the Celts she was worshipped as Cerridwen. Nor was she ignored in the New World; in what is now Mexico the Moon goddess was called Tlazolteotli and to the Mayans she was Ixchup. These names represent only a tiny proportion of those that are still remembered and there can be no doubt at all that Earth's Moon has been deeply important to humanity across the whole world and for many thousands of years.

The Moon was almost certainly the first heavenly body used to measure the passage of time for reasons other than human fertility. In this capacity it is still enshrined in our own systems by the use of months to split the solar year. Looking back at history it is easy to see the repeated attempts of different cultures to reconcile lunar time with a growing recognition of the length of the year, which is governed by the Sun. A truly ancient culture, such as that of the Sumerians, never abandoned its lunar calendar, beginning each month as the first crescent of the Moon showed itself in the dawn sky. However, at the same time Sumerian Priests adopted a 'stylized' month of thirty days in length, which fitted the solar year in a more regular way. Lunar reckoning is still used in Islam, a legacy of the religion's origins in the Arabian Peninsula.

In a physical sense this intense interest in the Moon is not at all surprising. We tend to forget in our modern world of electric lights that there was a time, not so long ago, when the Moon was a welcome sight on a dark night, but at the same time it was recognized to have awesome powers. It was believed by cultures from across the world that the Moon could have a bearing on people's mental states (see chapter five). The English word 'lunatic' enshrines this belief and, up to very recent times, it was considered that those who were mentally unstable could be triggered into madness and violence by the appearance of the full Moon. In addition, our ancient ancestors were well aware that the Moon was responsible for one of the most frightening and awe-inspiring happenings that periodically 'stole' the Sun from the sky.

Solar eclipses happen when the new Moon passes directly between the Sun and the Earth. At such times the shadow of the Moon is cast upon the Earth. If the observer is in the right place on the Earth, it will appear that the light of the Sun has been blotted out and day can suddenly become night. A total eclipse is a truly remarkable event because in order for it to happen the size of the Moon and the Sun, as seen from the Earth, must be identical. Nevertheless it does happen and it must have struck absolute terror into the hearts of early humans. This fear would have been slightly mitigated when it became possible to predict eclipses, something that a number of early cultures sought to do.

A second sort of eclipse, which is seen more often because of the planetary geometry involved, is called a 'lunar eclipse' – and in its own way this must have been just as potent and frightening. A lunar eclipse happens when the Moon moves through the shadow of the Earth, so the full Moon is seen to slowly disappear in a clear night sky. (See figure 19)

On these occasions the Moon's face is not totally blotted out by Earth's shadow, often appearing as a ghostly blood red disc. Even today this is a chilling sight and one can sympathize with people who viewed the event with a sense of foreboding.

Without a good understanding of the planetary cycles involved, eclipses of both sorts could easily appear to be random events and many early cultures sought to discover the patterns involved, probably working on the assumption that understanding inferred a degree of control. This may well have represented the first serious attempts at astronomy. It is known that both the Assyrians and the Babylonians could predict eclipses. In both cases many of the astronomical skills were inherited from the earlier Sumerians and it is highly likely that eclipse prediction already existed before 3,000BC.

Further west there have been suggestions that some Megalithic monuments were built as eclipse predictors, maybe as early as 4,000BC. Astronomer Gerald Hawkins in his book *Stonehenge Decoded* used a computer model to demonstrate that Stonehenge in Wiltshire, England, might have been partly built with eclipse prediction in mind.[5]

By at least the second millennium BC the Chinese could also predict eclipses. As far back as 2650BC, Li Shu was writing about the subject of astronomy. Three and a half centuries later, ancient Chinese astrologers had sophisticated observatory buildings, and solar eclipses were considered essential for forecasting the future health and successes of the emperor. These astronomers were keen to be accurate as failure to get the prediction correct was likely to be lethal for them. In one documented case referring to the eclipse of 2136BC the two astrologers who got it wrong were beheaded. The following recorded their fate: 'Here lie the bodies of Ho and Hi, Whose fate, though sad, is risible; Being slain because they could not spy Th' eclipse which was invisible.' – Author unknown

For thousands of years the Moon was a thing of awe and wonder to human beings across the entire planet and it remains so to millions of people today, despite technological advances and a good understanding of its physical characteristics. For example, the Moon has always been equated with agriculture. Even in some parts of the fully developed world there are farmers and gardeners who would not dream of either planting or harvesting without direct reference to the phase of the Moon or even the part of the zodiac it occupies at any particular point in time. The Moon is the fastest moving astronomical body when viewed from the Earth and appears to pass through all the zodiac signs in only 27.322 days.

Generally speaking, crops were often planted close to the new Moon, so that they could grow with the face of the Moon. Whilst there is no known scientific basis for such ideas, the advice offered is often very specific and doesn't vary much across the world. Nor does Moon-lore relate only to sowing seeds. For example, it is suggested that when picking apples for immediate eating, it is best to harvest them at the time of the full Moon, though if they are to be stored, the new Moon is preferred, since the apples are believed to be less likely to rot.

Even today the Moon has always been important to humanity and it is central to one of the most important festivals of the Christian religion. Easter, which falls in the early spring in the northern hemisphere, is an ancient celebration of rebirth that long predates its association with the death and resurrection of Jesus Christ.

The New Testament states that Jesus Christ was crucified on the eve of Passover before rising again a short time later. In consequence, the ancient Easter festival was reassigned to commemorate this miracle. There was, however, considerable debate over the date on which Easter should fall. The early Christians of Jewish origin celebrated the Resurrection immediately following their Passover festival, which, according to their lunar calendar, fell on the evening of the full Moon. This was the fourteenth day in the month of Nisan (the first month of their year), thereby causing Easter to fall on different days of the week. The new breed of non-Jewish Christians from around the Roman Empire wished to commemorate the Resurrection on a Sunday – their newly defined Sabbath. In 325AD the Roman emperor Constantine I convened the Council of Nicaea to debate whether or not Jesus Christ was a man or a god. Having officially designated Jesus to be God, by a narrow margin, the council then ruled that the Easter festival should be celebrated on the first Sunday after the full Moon following the vernal equinox; and that if the full Moon should occur on a Sunday and thereby coincide with the Passover festival, Easter should be commemorated on the Sunday following.

The origin of the word 'Easter' is thought to come from *Eostre,* the Anglo-Saxon name of a Teutonic goddess of spring and fertility. Her festival was celebrated on the day of the vernal equinox which now falls around March 21st when the Sun rises in the east and sets in the west, and the day has twelve hours of daylight and twelve hours of darkness. Traditions associated with this pagan festival survive in the idea of the Easter rabbit, a symbol of fertility, and in brightly decorated Easter eggs, which were a symbol of rebirth.

CHAPTER TWO

The Science of the Ancients

'The important thing is not to stop questioning. Curiosity has its own reason for existing.'

Albert Einstein

In the early 1930s a young Scottish engineer noticed that several of the widely ignored, prehistoric Megalithic sites near his home appeared to have lunar alignments. He decided to study some of the sites and he began a process of careful surveying that was eventually to lead him to make a discovery of staggering importance.

As a young engineer at Glasgow University, Alexander Thom visited a number of prehistoric stone structures near to his home in Scotland during the early 1930s. He marvelled at the grandeur and admired the way so many of the giant stones had survived the weathering of more than 5,000 years, as well as proving resistant to the thieving tendencies of croft and road builders across dozens of centuries. As he contemplated the various sites he mused over their purpose and as he looked to the horizon he could imagine how the stones might have been used as sighting stones for astronomical purposes. When he checked out the rising and setting points of the Sun and the Moon across the year his hunch appeared to be born out.

His first survey was at a site known as Callanish, on the Isle of Lewis in the Hebrides off the west coast of Scotland. This complex of standing stones revealed many astronomical alignments and is today often referred to as a 'Moon temple'. Thom went on to spend nearly half a century carefully surveying the so-called Megalithic (the word means giant stones) structures that lay scattered across the countryside from the islands off northern Scotland down to the French region of Brittany. Along the way he became a highly respected professor of Engineering at Oxford University until his retirement in 1961.

Thom had quickly realized that these prehistoric builders were engineers like himself and that they had a surprisingly sophisticated knowledge of geometry and astronomy. The approach taken by this talented engineer was to assess what he believed the site had been intended to do – and then redesign it himself. He quickly gained an empathy with the Stone-Age builders that gave him a real insight into the purpose of each site that would possibly be lost on a conventional archaeologist.

Once he had a picture in his mind of what he thought their plan had been, he went away to create his own solution to the assumed problem. Having drawn up his own design he then returned to compare the site layout to his own blueprint. Through this process he could predict the location of missing stones and, on further inspection, he would usually reveal the socket hole that confirmed his theory.

Thom developed a new statistical technique to establish the relative positions of the stones and, over time, something spectacularly unusual emerged from the amassed data. These prehistoric builders had not been lugging huge stones willy-nilly; they had manufactured these structures working with a standard unit of measurement across a huge area of thousands of square miles of what was then dense forest and barren moorland.

It was amazing that these supposedly primitive people could have had an 'international' convention for a unit of length, but the mystery deepens because Thom was eventually able to describe the supreme accuracy of a unit he called the Megalithic Yard. This was no approximate measure taken from paces or body parts; it was equal to 2.722 feet +/- 0.002 feet (82.96656cm +/- 0.061cm). Thom was also able to demonstrate that the unit was frequently used in its double and half form as well as being broken down into forty sub-units for use in design work that he designated as 'Megalithic Inches'.

Most archaeologists refuted the finding on the basis that the idea that a unit of measurement that was more accurate than a modern measuring tape was absurd. Thom admitted that he could not suggest how it could have been achieved but he stood by his evidence that simply said it 'had' been done. In our previous book, *Civilization One,* we described how we set out to investigate the concept of the Megalithic Yard. Our initial hypothesis was that if the unit was not an error of Thom's data analysis it logically should have two properties:

1 It should have an origin in something meaningful, rather than just being an abstraction that was adopted by everyone.
2 It should have a means of reproduction that could be used by anyone without reference to any sort of standard measuring rod, that would have been difficult to manufacture and impossible to keep accurate across centuries.

We realized that our assumption could be wrong on either or both counts but as it turned out, we were correct on both. Thom had not made an error.

As we describe in *Civilization One,* the Megalithic Yard is a geodetic unit, in that it is integral (has a whole number relationship) to the polar circumference of the Earth. We found that these early Megalithic builders viewed a circle as having 366 degrees rather than the 360 degrees that we use today. We realized that there really *should be* 366 degrees in a circle for the very good reason that there are 366 rotations of the Earth in one orbit of the Sun – the most fundamental of all circles in human existence.

One solar orbit is, of course, a year but there is a very slight difference between the number of rotations of the planet and the 365 days in a year. This is because the mean solar day is based on the time between the Sun being at its zenith on two consecutive days (86,400 seconds) but an actual rotation or 'sidereal day' takes 236 seconds less. All of those 'saved' seconds add up to exactly one more day over the year. A sidereal day can be easily appreciated by observing a star returning to the same point in the heavens on two consecutive nights. This is one spin of our planet because it is unaffected by the secondary motion of the Earth's orbit around the Sun.

Wheels within wheels

Early cultures frequently took their lead from nature and they were fond of 'wheels within wheels'. If the circle of the heavens had 366 parts, why should every circle not follow the same rule? We were able to confirm this hypothesis by a variety of means including evidence from later cultures that appear to have adopted the 366-degree principle.

The approach our Megalithic ancestors took, we argue, was to hypothetically divide the circle of the Earth into 366 degrees with sixty minutes per degree and six seconds per minute. It was reasonable to assume that these ancient builders used the polar circumference of the Earth that passed through the area around the British Isles. Our planet is nearly spherical but it does have a bulge in the centre between the poles, so the equatorial circumference is a little longer that the polar. There are varying estimations of the Earth's polar circumference, with NASA, for example, quoting an average figure of 39,941km, whilst other sources regularly quote 40,006km or 40,010km – but the most frequently used figure appears to be 40,008km. Undoubtedly much depends on where the measurement is taken or if an average of them all is calculated.

Interestingly, the shortest polar circumference (one that has least landmass) is the one that passes through the British Isles and is now considered as the zero line of longitude.

But there is also another possibility.

Just for interest, we looked at the flattest possible circumference achievable on the globe, i.e. a line that equally bisects the planet that has most sea and least land. We were amazed to discover that a person standing in the middle of Salisbury Plain in Wiltshire, England (where Stonehenge and the Megalithic circle at Avebury were built) is in the absolute centre of such a line. This means that if we consider Stonehenge to be the 'top' of the world, the imaginary equator from that point is almost 98per cent sea – more than any other point on Earth. This line passes across the South Atlantic, skims just below Africa, moves up across the Indian Ocean, clips small pieces of land at Banda Aceh, Sumatra, Thailand and Vietnam, over the South China Sea and then more than 20,000 kilometres across the Pacific to pass over a section of South America.

As far as we know such a line has not been measured, and we cannot imagine how it could have been measured without the aid of modern satellite technology. However, just because we do not know how it could have been done does not mean that it was not done. Without further evidence we have to assume that it is pure coincidence that Stonehenge stands on the only place on Earth to be equidistant from the optimum and near perfect sea-level circumference of the globe.

We can only assume that a polar circumference was used and taking the 40,008km figure it translates to 48,221,838 Megalithic Yards using Thom's central value for the unit. It was then subdivided as follows:

Polar circumference	=	48,221,838 MY
1 Degree (1/366th)	=	131,754 MY
1 Minute (1/60th)	=	2,196 MY
1 Second (1/6th)	=	366 MY

So, this brilliant system of geometry starts with 366 degrees and finishes with seconds of arc that are 366 Megalithic Yards long. Self-evidently, an amazing set of 'wheels within wheels'!

We knew that the system must work this way because we found that the later Minoan culture, which developed on the Mediterranean island of Crete around 2000BC, also used the Megalithic second of arc. However, the Minoans sub-divided it into 1,000 parts to become their standard unit of measure that was equal to 30.36cm. This unit was dubbed the 'Minoan Foot' by the Canadian archaeologist, Professor Joseph Graham who first detected its use in the palaces of ancient Crete.[6]

We went on to demonstrate how any person could generate a highly accurate Megalithic Yard by measuring the movement of Venus in the evening sky using a rope, some twine, a blob of clay, and a few sticks. The secret was to take one 366th part of the horizon and time the passage of Venus across it, and then to cause a piece of twine with a blob of clay on the end to swing like a pendulum 366 times during that period. From fulcrum to the centre of the blob was a mathematically perfect 1/2 Megalithic Yard or twenty Megalithic Inches. The process was simple to carry out and works on the fact that a pendulum is responsive to only two factors: the length of the pendulum and the mass of the Earth. If the pendulum beat 366 times during the transit of Venus across a 366th part of the sky – you had your measure! (See Appendix 1 for a more detailed explanation of the pendulum method.)

It is doubtful that these ancient stonemasons realized the fact but the period of time that they watched Venus and elected to subdivide into 366 beats, is equal to the difference between a mean solar day and a sidereal day.

Our starting point had been to search for all possible sources of reliable measurement available from nature. And we found that there was only one: the turning of the Earth on its axis as seen by watching the movement of the heavens. It was possible to time the passage of a star, or in this case the planet Venus, with reliable accuracy using a pendulum. The pendulum then turned a unit of time into a unit of length because the timed beat will always produce a fixed length – with tiny variations due to latitude and altitude.

It was then a simple matter to turn a unit of length into a measure of volume and capacity by creating cubes and filling them with liquid or dry goods such as barley or wheat. However, we were not prepared for the shock we received when we created a cube with sides of four Megalithic Inches and found that it held a pint that was accurate to a staggering one part in 5,000 against the standard laid down in the year 1601. Doubling the sides to eight Megalithic Inches produced an accurate gallon and doubling again produced the old dry measure known as a bushel. The mystery was compounded when we filled the 'pint' cube with barley and found that it weighed exactly one pound!

Things turned from the sublime to the ridiculous when further experimentation showed that a sphere with a diameter of six Megalithic Inches held virtually one litre and one ten times the size weighed a metric tonne when filled with water; all to an accuracy of better than 99 per cent.

The fact that Thom's apparently meaningless Megalithic Yard, extracted from surveying hundreds of prehistoric ruins, produces these cubic and spherical feats is not debatable. No one, no matter how sceptical they might be, can deny the simple maths. Neither can they deny that the odds of such compounded apparent connections being coincidence are very high. Yet, the pound and the pint are thought to be Medieval and the litre and the tonne were invented at the end of the eighteenth century.

A connection seemed impossible.

Then we looked at the Sumerian people who lived in the region we now call Iraq some 5,000 years ago. They are attributed with inventing writing, glass, the wheel, the hour, minute and second of time as well as the 360-degree circle with its subdivisions of 60 minutes and 60 seconds of arc. Quite amazing people.

As we probed the achievements of this civilization we found that the unit of length the Sumerians had used was virtually a metre at 99.88cm and that they had also used weights and capacities that were as equally matched to the kilo and litre of the French metric system created thousands of years later. Quite a coincidence we thought – but it was nothing of the kind, for when we applied the principles of the pendulum to the Sumerian unit of length called the 'double kush' we found that a pendulum of this length beat at the rate of one per second. This meant that the Sumerian's key unit of length and their key unit of time were two sides of the same coin when used as a pendulum. A double-kush pendulum would always beat out a second and a pendulum that beat at the rate of a second would always be a double kush in length. This demonstrates beyond all reasonable doubt that the Sumerians used pendulums to define their measurements. The question was, had they used the same Venus-watching principle as the Megalithic builders of the British Isles to reproduce their units?

Sumerian written records tell us that the planet Venus was considered to be the goddess Inanna, who was of central importance to their culture, so it seemed entirely plausible. If they had used the same principle it seemed logical that they would have employed their own values; essentially keeping the same 'software' but inputting their own data. Instead of the 366 degrees of the Megalithic system we would have to use the more familiar 360 degrees first used by the Sumerians. And when we checked out the results for such a process – it worked perfectly.

When the horizon was divided into 360 parts and Venus was timed across that part of the sky at the appropriate time of year the double-kush pendulum metres out exactly 240 seconds. And the period of 240 seconds is recorded as being so important to the Sumerians it had its own name – a 'gesh'. It therefore seems certain that these people followed the Megalithic idea of creating a unit of length from timing the movement of Venus across the evening sky.

The American connection

Later in our research we came across a letter written by the great American statesman, Thomas Jefferson and sent to the House of Representatives on July 4th 1776. In this letter

Jefferson laid out a recommendation for a new system of weights and measures for the new United States that he had helped to establish. He gave his reasoning and described some unusual facts he had uncovered whilst developing his intended units.

He explained how he had realized that there was only one aspect of nature that gave rise to any reliable unit of measure – which he named as the turning of the Earth. So, like ourselves and the Megalithic builders of five and six millennia before him, he used the heavens to provide a basis for all measurement. In his letter he stated that he had come to realize that the imperial system of measurement used in Britain was not an accumulation of unrelated units as generally imagined. On the contrary, he said that their harmony indicated to him that they were members of a group of measurement units 'from very high antiquity'.

He gave a number of reasons for this belief including his astonishment that the foot, made up of twelve inches, was directly related to the ounce weight through the use of cubes. He said: 'It has been found by accurate experiments that a cubic foot of rain water weighs 1000 ounces avoirdupois (Imperial).'

It could be coincidence that a cubic foot holds 1,000 ounces of rainwater, not 999 or 1,001, but exactly 1,000 – or that the cube has sides that are a perfect 10x10x10 one-tenths of a foot. But Jefferson did not think so. And nor do we. However, it was Jefferson's proposed units that fascinated us. They were never adopted but their properties are amazing.

Jefferson's logical mind also caused him to use a pendulum to convert time into a linear unit. He decided that he should use a pendulum that had a beat of one second as the basis for his measuring system. Of course, Jefferson had no idea that the second had come from the Sumerian culture or that it had been created by the use of a pendulum in the first place. Jefferson added one improvement suggested to him by a certain Mr Graham of Philadelphia – that he use a rigid pendulum of very thin metal without a weight on the end because it is more accurate than a conventional type of pendulum. The rules change with such a pendulum (known as a rod). A rod has to be exactly 50 per cent longer than a pendulum to produce the same time period. Jefferson's timing piece, that beat once per second, is known as a 'seconds rod', and is 149.158145cm in length.

The world knew nothing of the Sumerian culture in Jefferson's time and he could not possibly have been aware that his rod that beat once per second was essentially three kush in length – just a whisker less than one and a half metres (remembering that the metre had not been invented at that time).

The three-kush rod behaves exactly like a double-kush pendulum and therefore it beats 240 times during one 360th part of a day;

observable by watching Venus move across a 360th part of the sky. Jefferson was therefore accidentally re-enacting the ritual used by Sumerian astronomer priests nearly 5,000 years earlier and connecting with the principles of prehistoric measurements.

The units that Jefferson identified from this ancient process were all based on the length of this 'seconds rod'. He wrote:

> 'Let the second rod, then, as before described, be the standard of measure; and let it be divided into five equal parts, each of which shall be called a foot; for, perhaps, it may be better generally to retain the name of the nearest present measure, where one is tolerably near. It will be about one quarter of an inch shorter than the present foot.
> Let the foot be divided into 10 inches;
> The inch into 10 lines;
> The line into 10 points;
> Let 10 feet make a decad;
> 10 decads one rood;
> 10 roods a furlong;
> 10 furlongs a mile.'

We can see that his proposed 'decad' was based on a double-seconds rod. It was equivalent to six Sumerian kush, and his furlong was equal to 600 kush. This brings about an even deeper connection with the people of ancient Iraq because they used a system of counting that was sexagesimal; which means it used a combination of base ten and base sixty. They had a system of notation that worked as follows:

Step	multiple		Value
1.	1	=	1
2.	x 10	=	10
3.	x 6	=	60
4.	x 10	=	600
5.	x 6	=	3,600

It can be seen that the figure of 600 is indeed a Sumerian value for a Sumerian unit of length.

But not only is the Jefferson furlong equal to 600 kush – it is also an almost perfect 360 Megalithic Yards.

Strangely, Jefferson had connected well with both the Megalithic and the Sumerian system. But something even stranger happened when we took Jefferson's furlong and multiplied it by 366 and 366 again:

$$366^2 \text{ furlongs} \quad = \quad 39{,}961.257\text{km}$$

As we have already mentioned, the range of assumed lengths of the Earth circumference varies by a few kilometres depending on what source one consults, probably because each cross section will differ and tides and plate tectonics involving mountains leave room for some debate. At the higher end 40,008 kilometres is widely used, however if we take NASA preferred figures they quote a polar radius of 6,356.8 kilometres which equates to a polar circumference of 39,941.0 kilometres.

That means that 366^2 Jefferson furlongs match Nasa's estimate of the Earth's size to an accuracy of 99.95 per cent – which is as perfect as it gets!

Problems with Foucault's pendulum

We became more and more fascinated by everything to do with pendulums. During one particular telephone conversation, which had gone on for over an hour, we had, yet again, discussed at length the idea that there might be some unknown law of astrophysics – that was revealed by pendulums – at work here. We considered some highly speculative thoughts that ranged from standing electromagnetic sine waves due to a gyroscopic effect of the Earth's spin through to gravitons containing packets of information about 'geometrical shape'. But we agreed that we just did not know enough to even start to investigate such ideas. Chris wrote the following paragraph into a draft of this chapter as a summary of our mutual frustration and finished work for the day.

'We have to admit that we still do not understand why it is so, but the use of pendulums in association with these ancient values appears to be elemental to the planet Earth – some physical reality seems to be at work here. Every pendulum reacts to the mass of the Earth but there seems to be some kind of 'harmonic' response at certain rhythms: points where the mass and the spin of the planet resonate in some way.'

But at that very point in time everything changed.

At five o' clock the following morning Chris was unable to sleep and decided to get up and make a cup of tea. It was then that a 'library angel' turned up.[7] Looking for something to read he pulled the delivery sleeve of a magazine that had arrived in the post the previous day and flicked it open. The main feature article in this edition of *New Scientist* was entitled: 'Shadow over gravity'. It sounded interesting even early on a dark November morning.

But he quickly realized it was far more important than merely 'interesting'. The opening paragraph was incredibly similar to that which opens this book, carrying a description of how it feels to witness a total eclipse – and then it transpired that the thrust of the article was that solar eclipses have a profound effect on pendulums! A debate is presently raging as to why this should be the case, because the suggestion has been made that pendulums may well be the key to a significant hole in Einstein's theory of relativity.

The starting point concerns the work of Jean Bernard Leon Foucault who demonstrated a special quality of pendulums at the Great Exhibition, held in London in 1851. His pendulum, now always referred to as 'Foucault's pendulum', is simply a very heavy weight fastened to a very long wire attached to a ceiling inside a very tall building, with a universal joint allowing it to rotate freely around a fixed point so that it will swing in a slow arc in any direction. Giant pendulums of this kind are now routine exhibits at some of the major museums around the world including the Smithsonian in Washington and the Science Museum in London.

Once set in motion its direction of swing will appear to rotate at a rate of about twelve degrees an hour. But this is actually an illusion because it is the observer and the rest of the world that is moving whilst the pendulum is maintaining a fixed swing back and forth in relation to the Universe. This happens because the pendulum is independent of the movement of the Earth, which is rotating underneath the pendulum, making it appear that the pendulum is changing direction. The reason a pendulum swings is because the Earth's gravity continually tugs down on it. According to Einstein's general theory of relativity this relentless tugging is due to the fact that every mass bends the fabric of space-time around it causing other masses to slide down into the dimple it creates in space-time.

The amount of rotation of a Foucault pendulum is dependent on latitude. At the North or South Pole the pendulum appears to rotate through an entire 360 degrees once every turn of the Earth (each sidereal day) because the planet rotates all the way round underneath it. In the northern hemisphere at the latitude of the British Isles the rate of rotation is reduced to around 280 degrees per day and the rate of rotation continues to fall the closer one gets to the equator, where a Foucault pendulum does not rotate at all.

For over a hundred years everyone knew that a Foucault's pendulum would swing in an entirely predictable manner at any specific location. Then in 1954 a French engineer, economist and would-be physicist by the name of Maurice Allais found that this was not always the case. He was conducting an experiment at the School of Mining in Paris to investigate a possible link between magnetism and gravitation, in which he released a Foucault pendulum every fourteen minutes for thirty days and nights, recording the direction of rotation in degrees. By chance, a total solar eclipse occurred on one of those days.

Each day the pendulum moved with mechanical precision but on June 30th 1954, when a partial eclipse occurred, one of Allais' assistants realized that the pendulum had gone haywire. As the eclipse began, the swing plane of the pendulum suddenly started to rotate backwards. It veered furthest off course twenty minutes before maximum eclipse, when the Moon covered a large portion of the Sun's surface before returning to its normal swing once the eclipse was over. It seemed that the pendulum had somehow been influenced by the alignment of the Earth, the Moon and the Sun.

This was totally unexpected and utterly startling. Allais' experiment was being conducted indoors, out of the sunlight so there was no apparent way the eclipse could have affected it. Allais was at a loss to explain what had taken place but when he conducted an improved version of his experiment in June and July 1958 with two pendulums six kilometres apart he found the same effect. Then during the partial solar eclipse of October 22nd 1959, Allais once again witnessed the same erratic rotation – but this time similar effects were reported by three Romanian scientists who knew nothing of Allais' work.

Many people have questioned his results, mainly because science does not like that which it cannot explain. Many others have now repeated the experiment with mixed results: some found no measurable effect, but most have confirmed the result at different locations – including one conducted in an underground laboratory![8]

It is interesting to note that in 1988 Allais was awarded a Nobel Prize for economics. Like Alexander Thom (and many other paradigm busters) a major discovery had come from someone working outside their own field. These are bright people who are driven by curiosity and who are not the products of conventional training.

Allais despairs at the standards of those that oppose without logic or reasoning: 'In the history of science, every revolutionary result meets with very strong opposition ... Relativists say I'm wrong without providing any demonstration. Most of them haven't even read what I wrote.'

In 1970 Erwin Saxl and Mildred Allen of Mount Holyoke College, Massachusetts, studied the behaviour of a pendulum before, during and after a total eclipse. The pair took a slightly different approach to Allais as they used a torsion pendulum, which is a massive disc suspended from a wire attached to its centre. Rotating the disc slightly causes the wire to twist. When it is released, the disc continues to twirl first clockwise, then anticlockwise, with a fixed period. But during an eclipse, their pendulum sped up significantly. They concluded that gravitational theory needs to be modified.

In India in 1995, D C Mishra and M B S Rao of the National Geophysical Research Institute in Hyderabad observed a small but sudden drop in the strength of gravity when using an extremely accurate gravimeter during a solar eclipse. But results have been mixed. When the eclipsed Sun rose above Helsinki on July 22nd 1990, Finnish geophysicists found no disturbance to the usual swing, yet in March 1997 scientists observed gravimeter anomalies during an eclipse in a very remote area of north-east China.

The mystery continues and yet no academic institution appears willing to invest time and money to study this phenomenon in depth. However, Thomas Goodey, a self-funding independent researcher from Brentford in England, has decided that he will investigate the Allais effect by using several pendulums during an eclipse. Because modern equipment is much more accurate and sensitive than that available in 1954 – giving twenty to one hundred times better resolution, he is confident of a clear result.

Goodey plans to travel the world over the next few years with twelve specially constructed pendulums. In May 2004, he presented his strategy at a meeting of the Society for Scientific Exploration in Las Vegas and invited physicists to join him. As *New Scientist* reported, several leapt at the chance.

Goodey suspects that the anomalies occur when an observer is near the line that connects the centres of masses of the Sun and the Moon. During a total solar eclipse, the Sun–Moon line intersects the surface of the Earth at two points on roughly opposite sides of the globe. This theory would explain why the sunrise eclipse in Helsinki did not produce a result. Goodey is quoted as saying that observations at this 'anti-eclipse' point where no eclipse is visible might carry much greater weight.

We wait with interest to hear the final results of Thomas Goodey's experiments. At this point it seems as though we might well have been right to suspect that pendulums reveal a great deal about the nature of our planet's gravity and its gravitational relationship with the Moon and the Sun. Could it be that because the Moon blocks out the disc of the Sun so perfectly it is acting as a shield to an ongoing interaction between the Earth and the Sun? Or perhaps it is because all three centres of mass are lined up and something physical occurs at this time?

We also wonder whether the unknown individuals who devised the Megalithic Yard and its inherent geometry understood much more about this pendulum effect than we do. Our previous findings strongly suggest that they knew a great deal more about the Earth –Moon–Sun relationship.

A special relationship

Our initial findings about Megalithic geometry, described in *Civilization One,* had caused us to examine all kinds of unexpected relationships between the Earth and ancient measures. This had further prompted us to wonder whether the 366 geometry, that produced the Megalithic Yard, was in some way planet specific. Was there some connection between the mass, spin and solar orbit that made it special to the Earth?

First we applied the principles of Megalithic geometry to all of the planets of the solar system. No discernable pattern emerged – they appeared to be completely random results. For example Mars produced 19.78 Megalithic Yards per second of arc and Venus an unimpressive 347.8. We also checked out the major moons of other planets to no avail.

A good friend of Chris, Dr Hilary Newbigen, suggested that, for thoroughness, we try using the number of days per orbit for each planet to see if there was a relationship to the individual dimensions, but again the results were negative.

Then we looked at Earth's Moon.

The result here was anything but meaningless. We took the Moon's radius, defined by NASA as being 1,738,100 kilometres, to calculate a circumference of a meaningless sounding 10,920,800 metres. We then converted this distance into Megalithic Yards, which gave us the equally apparently arbitrary value of 13,162,900.

We then applied the rules of Megalithic geometry by dividing this circumference into 366 degrees, sixty minutes and six seconds of arc. To our total amazement there were 100 Megalithic Yards per lunar Megalithic second of arc. The accuracy of the result was 99.9 per cent which is well within the range of error of this kind of calculation.

How strange that the Megalithic Yard is so elegantly 'lunardetic' as well as geodetic!

Our next thought was the Sun. Because we know that the Sun is 400 times the size of the Moon it should logically have a perfect 40,000 Megalithic Yards per second of arc. For thoroughness we checked out the sums and it did indeed work as perfectly as we expected.

This all seemed very odd. The Megalithic structures that were built across western Europe were frequently used to observe the movements of the Sun and the Moon, but how could the unit of measure upon which these structures were based be so beautifully integer to the circumference of these bodies as well as of the Earth?

Is it coincidence? On top of all the other strange facts regarding the Moon it becomes rather unrealistic to keep putting everything down to a random fluke of nature. Of course, we were well aware that the numbers we were looking at were only integer when one uses base ten – and we will deal with that issue later.

If it is not coincidence then there are only two other options. The first is that there is some unknown law of astrophysics at work, causing relationships to emerge that were spotted in some way by our Stone-Age forebears. The other is conscious design.

The idea of deliberate design seemed plum crazy – common sense tells us it's wrong. Then we, once again, considered more wise words from Albert Einstein: 'Common sense is the collection of prejudices acquired by age eighteen.'

At the age of eighteen we, like everyone else, 'knew' that everything in the world was natural. But when we put our prejudices of what can and cannot be, to one side and thought laterally about it, the more reasonable it seemed.

It was not unreasonable to believe that the stonemasons of the Neolithic period were smart enough to measure the polar circumference of the Earth and that they devised a unit of measure that was integer to the planet. Such a feat can be achieved with very simple tools as demonstrated by the Ancient Greeks. But could they really have measured the circumference of the Moon and the Sun?

Or was this mysterious property of pendulums something to do with it?

Most of all we marvelled at the fact that, yet again, it was the size and position of the Moon that revealed that there is an issue to resolve.

CHAPTER THREE

The Origin of the Moon

'The best explanation for the Moon is observational error – the Moon does not exist!'

Attributed to Irwin Shapiro of The Harvard-Smithsonian Center for Astrophysics

The one inescapable fact about the Moon is that it orbits the Earth. It is up there beaming down on us, but according to everything that science knows, it shouldn't be.

As we have seen, it is known that people have been Moon-gazing for tens of thousands of years, and our understanding has grown to a point where we are now very confused.

The Greeks were great gatherers of knowledge and investigators of the rules of nature. In the fifth century BC Democritus, who originated the theory that matter was made of indivisible units he called atoms, went to the other end of the scale and suggested that the markings on the Moon could be mountains. A little later Eudoxus of Cnidus, who was an astronomer and mathematician, calculated the

Saros cycle of eclipses and thereby could predict when they would appear.

Around 260BC, yet another Greek by the name of Aristarchus, devised a method by which he thought he could measure the size of the Moon and gauge its distance from Earth. He never actually achieved it but a mathematician and astronomer of major importance known as Hipparchus of Rhodes achieved the feat around a hundred years later. He used an ingenious technique that was conducted during a solar eclipse. The eclipse in question was total in Syene but only partial in Alexandria which was some 729 kilometres away. Enlisting the help of like-minded friends, Hipparchus was able to use the known distance from Syene to Alexandria, together with the angular difference of the total and partial eclipse to establish the Moon's true size and distance from the Earth.

At the end of the first century AD, Plutarch wrote a short work about the Moon, entitled *On the Face in the Moon's Orb* where he suggested that the markings on the face of the Moon were caused by deep recesses, too deep to reflect sunlight. He proposed that the Moon had mountains and river valleys and even speculated that people might live there.

Although a Hindu astronomer, Aryabbata, repeated and confirmed the experiment conducted by Hipparchus as late as 500AD, Christian authorities of the time maintained a biblical approach to the Moon and only information about our near neighbour that didn't contradict the scriptures was countenanced. With the arrival of Christianity the world entered a dark age where scripture rather than science was the only permitted guide to human existence.

The grip of the Church slipped somewhat during the fifteenth and sixteenth centuries and the Renaissance (literally meaning 'rebirth') emerged bringing radical and comprehensive changes to European culture. The Renaissance brought about the demise of the Middle Ages and for the first time the values of the modern world made an appearance. The consciousness of cultural rebirth was itself a characteristic of the Renaissance. Italian scholars and critics of this period proclaimed that their age had progressed beyond the barbarism of the past and had found its inspiration, and its closest parallel, in the civilizations of ancient Greece and Rome. By the end of the sixteenth century, a genius from the town of Pisa called Galileo Galilei became one of the most important scientists of the Renaissance carrying out experiments into pendulums, falling weights, the behaviour of light and many other subjects that captured his imagination. Above all, for most of his adult life Galileo was an avid astronomer.

In May 1609, Galileo received a letter from Paolo Sarpi telling him about an ingenious spyglass that a Dutchman had shown in Venice. Galileo wrote in April 1610:

'About ten months ago a report reached my ears that a certain Fleming had constructed a spyglass by means of which visible objects, though very distant from the eye of the observer, were distinctly seen as if nearby. Of this truly remarkable effect several experiences were related, to which some persons believed while others denied them. A few days later the report was confirmed by a letter I received from a Frenchman in Paris, Jacques Badovere, which caused me to apply myself wholeheartedly to investigate means by which I might arrive at the invention of a similar instrument. This I did soon afterwards, my basis being the doctrine of refraction.'

From these reports, and by applying his skills as a mathematician and a craftsman, Galileo began to make a series of telescopes with an optical performance much better than that of the Dutch instrument. His first telescope was made from available lenses and gave a magnification of about four times, but to improve on this Galileo taught himself to grind and polish his own lenses and by August 1609 he had an instrument with a magnification of around eight or nine. He quickly realized the commercial and military value of his super-telescope that he called a *perspicillum,* particularly for seafaring purposes. As the winter of 1609 brought colder, clearer nights Galileo turned his telescope towards the night sky and began to make a series of truly remarkable discoveries.

The astronomical discoveries he made with his telescopes were described in a short book called *The Starry Messenger* published in Venice in May of the following year – and they caused a sensation! Amongst many other findings Galileo claimed to have proved that the Milky Way was made up of tiny stars, to have seen four small moons orbiting Jupiter and to have seen mountains on the Moon.

As with many of his scientific investigations Galileo could easily have fallen foul of the Church authorities if his drawings of the Moon had been made public. According to Christian tradition both the Sun and the Moon were perfect, unblemished spheres. They simply had to be so because God had created them – and none of the Almighty's creations could be flawed. Eventually Galileo was put under perpetual house arrest by the Papacy for his blasphemous claim that the Sun was at the centre of the solar system. It is therefore quite possible that he knew more about the Moon than he was willing to admit in public.

In order to explain the markings on the Moon without treading on the toes of the Church, a number of ideas were put forward in Christian countries. Perhaps the most popular of these, at least for a while, was the suggestion that the Moon was a perfect mirror. If this was the case there were no markings on the Moon but rather reflections of surface features on the Earth. It didn't seem to occur to anyone that as the Moon orbited the Earth the markings should change, since the land beneath it would not remain constant.

Another suggestion, and one that was accepted in some circles, was that there were mysterious vapours between the Earth and the Moon. The images, it was suggested, were present in sunlight and were merely being reflected from 'the vapours'. But the most popular theory, probably because it didn't impinge on Christian doctrine, was that there were variations in the density of the Moon and that these created the optical illusions we see as markings on the Moon's surface. This unlikely explanation was safe, though it probably did little to convince early scientists, and certainly would not have impressed Galileo.

After Galileo's time, telescopes improved markedly and it was obvious to anyone who studied the Moon that it was a sphere with a rocky and undulating surface. As the Church gradually lost its power to direct scientific thought, many of the earlier ideas regarding the Moon became unthinkable. But no one had any idea how the Moon had come into being and why it occupied the orbit it did around the Earth.

It didn't take long for the subject of the Moon to become very important to astronomers. Empires such as those created by Britain, France and Spain, were expanding. This necessitated long sea voyages and led to that most urgent of searches – a way to plot 'longitude' whilst at sea. It is quite easy to establish one's position on the planet in a north–south line (latitude) but it was impossible to know where you were in terms of east–west (longitude). In the northern hemisphere, for example, latitude can be quickly gauged by measuring the angular distance between the horizon and the Pole Star. This angle also defines one's position north of the equator.

The longitude problem was eventually solved by having an extremely accurate clock on board a ship that was set to the time at one's point of departure. It wasn't difficult to work out the difference between local time, say at midday, and the time at the home port. It was then simply a matter of adding or subtracting to discover one's true position on the Earth's surface. This was fine but it took many decades before a suitably accurate clock could be created. In the meantime, astronomers sought for other methods to determine longitude, not least of all because there was a fabulous prize on offer for anyone who could crack the problem. And the place where many of them turned to establish longitude was the Moon.

Astronomers proposed that if really accurate tables were kept of the Moon's position relative to the background stars it would be possible to assess the true time of day in one's home port. The reason this could work was that the Moon, being very close to the Earth and orbiting quickly, moved across the heavens by around thirteen degrees of arc per day. Using the Moon it was a fairly simple matter to establish 'local time' and then to do the necessary computations to discover one's position.

The lists of tables necessary to accomplish the task were not so simple, however, and as soon as good chronometers were available the Moon was abandoned as a means for longitude assessment. However, the desire to solve this problem, and the potential profitability of doing so, meant that the Moon was receiving a great deal of attention during the seventeenth century and very accurate maps of its surface began to appear.

It wasn't until the nineteenth century, however, that probably the first reasoned explanation as to the Moon's origin was put forward. George Darwin, the son of Charles Darwin, the controversial Englishman who first proposed the theory of natural selection, was a known and respected astronomer who studied the Moon extensively and came up with what became known as the 'fission theory' in 1878. George Darwin may have been the first astronomer to ascertain that the Moon was moving away from the Earth. Working backwards from his knowledge of the rate the Moon was receding from the Earth, Darwin proposed a time that the Earth and the Moon could have been part of the same common mass. He suggested that this molten, viscous sphere had been rotating extremely rapidly in about five and a half hours.

Darwin further speculated that the tidal action of the Sun had caused what he termed as 'fission' – a Moon-sized dollop of the molten Earth spinning away from the main mass and eventually taking up station in orbit. At the time this seemed very reasonable and was the favoured theory by the beginning of the twentieth century. In fact the fission theory did not come under serious attack until the 1920s when a British astronomer called Harold Jeffries was able to show that the viscosity of the Earth in its semi-molten state would have dampened the motions required to generate the right sort of vibration necessary to fulfil Darwin's fission.

A second theory that once convinced a number of experts was the 'coaccretion theory'. This postulates that the Earth, having already been formed, accumulated a disc of solid particles – a little like the rings of Saturn. It was suggested that, in the case of the Earth, this disc of particles ultimately came together to form the Moon. There are several reasons why this theory can't be the answer. Not least is the problem of the angular momentum of the Earth–Moon system that could never have been as it is, if the Moon had formed in this way. There are also difficulties regarding the melting of the magma ocean of the infant Moon.

The third theory regarding the origin of the Moon that was in circulation around the time that the first lunar probes were launched was the 'intact capture theory'. At one time seeming to be the most attractive possibility, the intact capture theory suggested that the Moon originated far from the Earth and that the Moon became a 'rogue' body that was simply captured by the gravitational pull of the Earth and that it took up orbit around the Earth.

There are many reasons why the intact capture theory is now disregarded. Oxygen isotopes of the rocks on the Moon and on the Earth prove conclusively that they originated at the same distance from the Sun, which could not be the case if the Moon had been formed elsewhere. There are also insurmountable problems in trying to build a model that would allow a body as big as the Moon to take up orbit around the Earth. Such a huge object could not simply drift neatly into an Earth orbit at low speed like carefully docking a supertanker – it would almost certainly smash into the Earth at a massive speed or possibly skim off and hurtle onward.

By the middle of the 1970s all previous theories about the way the Moon had been formed were running into trouble for one reason or another and this created a virtually unthinkable situation in which acclaimed experts might have to stand up in public and admit that they simply didn't know how or why the Moon was there. As acclaimed science writer William K Hartmann, senior scientist at the Planetary Science Institute, Tucson, Arizona said in 1986 in his book *Origin of the Moon:*

> 'Neither the Apollo astronauts, the Luna vehicles, nor all the king's horses and all the king's men could assemble enough data to explain the circumstances of the moon's birth.'[9]

Out of this miasma came a new theory and, in fact, the only one that is presently widely accepted despite some fundamental problems. It is known as the 'Big Whack theory'.

The idea came out of theories that originated in the Soviet Union in the 1960s – specifically the work of Russian scientist V S Savronov, who had been working on the possibility of planetary origins from literally millions of different-sized asteroids known as planetesimals.

As a divergence from the Soviet ideas, Hartmann, together with a colleague, D R Davis, suggested that the Moon had come into being as a result of the collision of two planetary bodies, one being the Earth and the other a rogue planet at least as large as the planet Mars. Hartmann and Davis hypothesized that the two planets had collided in a very specific way that allowed jets of matter to be ejected from the mantles of both bodies. This matter was thrown into orbit, where it eventually came together to form the Moon.[10]

The suggestion seems to have many merits. First and foremost it appears to address the greatest puzzle that the recovery of Moon rock had thrown up: How was it that the composition of the Moon was so similar to that of the Earth, but only in part?

A close analysis of Moon rock has shown that it is very similar to the rock that forms the mantle of the Earth, yet the Moon is nowhere near as massive as the Earth in proportional terms. (The Earth is only 3.66 times as big as the Moon but has eighty-one times the Moon's mass.) It was obvious that the Moon could not contain many of the heavy elements that are found inside the Earth and the Big Whack theory purported to explain why this was the case. The Earth and the rogue visitor had come together in a very specific way. Although they would eventually form one planet it was reasoned that they must have impacted, drawn apart and then come together again. Computer modelling showed that under these very special circumstances it would have been possible for the material thrown off to have been mantle material, from close to the surface of the two bodies.

Although the theory eventually gained ground, at first it seemed so improbable that it was generally rejected. But with the passing of time, further work showed that such an unlikely scenario could conceivably have taken place. In 1983 an international conference was held at Kona, Hawaii, to try and solve the problems regarding the origins of the Moon. It was at this meeting that the Big Whack theory, also known as the Giant Impact Hypothesis of the Collision Ejection theory, began to gain ground. Hartmann's own suggestions, together with those of other scientists at the conference, formed the nucleus of the 1986 book, *Origin of the Moon,* which was edited by Hartmann himself.

In the intervening period several experts have created computer models that purport to add weight to the Big Whack theory and the most convincing of these are those of Dr Robin Canup, who is now Assistant Director of the Department of Space Studies in Colorado, USA. Canup wrote her PhD dissertation on the Moon's origin and specifically the Big Whack theory. Her early work led to the conclusion that the suggested impact would have actually led to a swarm of moonlets, rather than the Moon, but by 1997 further computer modelling resulted in a model of the impact that would lead to the Moon's presence.

Despite the fact that the Big Whack theory is now generally accepted by most authorities, it has many problems. Not least of all is that recognized by Robin Canup herself as she admits that there is one key aspect of the theory that doesn't make sense. This stems from the fact that other researchers have pointed out that such a massive impact as that proposed could not have failed to speed up the rotation of the Earth to a level far beyond today's situation. Canup agrees and the only way that she could deal with this anomaly is to propose a second major impact – which was designated 'Big Whack II'. This suggests that the second planetary collision happened perhaps only a few thousand years after the first one but, quite incredibly, this incoming object came from the opposite direction and so cancelled out the huge spin imparted to the Earth by the first cataclysmic event. This balanced double act sounds unlikely in the extreme. Two cosmic collisions that just happen to precisely return the planet to its natural rhythm? To us, this explanation smacks of desperation!

Canup herself is not happy with Big Whack II and is hopeful of modifying the original theory so that it can account for the present rate of spin of the Earth.

There is another big problem to overcome if the Big Whack theory is to be taken seriously. When rocks were brought back from the Moon, both by American astronauts and Soviet unmanned Moon missions, they were subjected to every conceivable test. The observed fact that put paid to the captured asteroid theory of the Moon is also a gigantic stumbling block to the Big Whack theory. It has been observed that the oxygen isotope signatures of Moon rocks are identical with those of rocks from the Earth – and that fact has some serious implications: Moon rocks and Earth rocks can only have the same oxygen isotope signature if they originated at the same distance from the Sun. This would mean that the Mars-sized body that hit the Earth must have occupied a similar orbit to that of the Earth and yet had already managed to survive for many millions of years before it hit the Earth.

That does not sound reasonable.

This situation is extremely unlikely and it throws up other difficulties. The present obliquity of the Earth (its twenty-three degree tilt against the plane of its orbit around the Sun) is usually deemed to be the result of the giant impact, but any body of the size of Mars that was in an orbit similar to that of the Earth could not have had sufficient momentum to knock the Earth's angle of rotation back so severely. Either the rogue planet was Mars-sized, and came from way out in the solar system and was therefore travelling extremely fast, or else it had to be at least three times the size of Mars, which doesn't tie in with the computer models as they stand.

Some of the other problems were cited by Jack J Lissauer, a well-respected scientist from NASA's Ames Research Center in an article he wrote for *Nature* in1997.[11] Lissauer is said to have joked to his students about a remark made by another scientist, Irwin Shapiro from the Harvard-Smithsonian Center for Astrophysics: 'The best explanation for the Moon is observational error – the Moon does not exist!'

Lissauer's article pointed out some of the problems with the Big Whack theory. He made it clear that in his opinion the latest research demonstrated that much of the material blown out by the impact (the ejecta) would have fallen back to the Earth. He says:

> 'The implication here is that lunar growth in an impact-produced disk is not very efficient. So, to form our Moon, more material must be placed in orbit at a greater distance from Earth than was previously believed.'

Lissauer made it clear that as a result, he too is of the opinion that the rogue planet must have been substantially larger than that originally proposed but noted that it is difficult to see how the excess angular momentum resulting from such a large impact could have been lost.

Three other scientists, Ruzicka, Snyder and Taylor, approached the problem from a slightly different direction by analysing the biochemical data available against the theoretical Big Whack. After a detailed examination they concluded: 'There is no strong geochemical support for either the Giant Impact or Impact-triggered Fission hypotheses.'[12]

These words used in the conclusion to this biochemical analysis indicate just how hopelessly contrived the whole Big Whack theory is. They go on to say: 'This [hypothesis] has arisen not so much because of the merits of [its] theory as because of the apparent dynamical or geochemical short-comings of other theories.'

In other words scientists hang onto the Big Whack theory, even though it has more holes than a rusty colander, simply because no other logical explanation has been found. It is just the least impossible explanation for a celestial body that has no right to be there.

Not only is the Big Whack theory discredited on a number of grounds by the scientific fraternity itself, it also singularly fails to deal with the anomalies thrown up by our own research, as outlined throughout this book. Big Whack cannot explain the extraordinary ratio relationship between the Moon and the Sun or the Moon and the Earth. The Moon could, by pure chance, end up being exactly 1/400th the size of the Sun and occupying an orbit that allows it to stand 1/400th the distance between the Earth and the Sun – but the odds are, quite literally, astronomically against it.

The Moon is proportionally bigger in relation to its host planet than any other in the solar system apart from Charon, Pluto's moon, which is more than half the diameter of Pluto. These two bodies are essentially twin planets or may be asteroids orbiting each other at close range although they are believed to have an unrelated origin.

Mercury has no moons at all and neither has Venus. Mars does have two moons but they are tiny in comparison with our own.

A close examination of the many samples of Moon rock brought back by the American Apollo missions and the Soviet unmanned missions has thrown up what turned out to be one of the biggest surprises of all. It has been observed that the oldest of the rocks collected from the Moon are significantly more ancient that any rock ever found on Earth. The most venerable rocks to be found on the Earth date back 3.5 billion years, whilst some samples from the Moon are around 4.5 billion years old – which is very close to the estimated age of our solar system. When radioactive dating techniques are applied to meteorites they are uniformly found to be 4.6 billion years old.

Yet even these rocks have the same oxygen isotope signature as those on Earth, another indication that the Moon has occupied its present distance from the Sun for an incredibly long time. There is currently no persuasive argument for this state of affairs.

Our own, almost accidental, discoveries regarding the peculiar ratio relationships between the Earth, Moon and Sun described in our previous book, *Civilization One,*[13] led us to an in-depth appraisal of the latest theories regarding the Moon and its origins. We were stunned by what we discovered. The Moon is bigger than it should be, apparently older than it should be and much lighter in mass than it should be. It occupies an unlikely orbit and is so extraordinary that all existing explanations for its presence are fraught with difficulties and none of them could be considered remotely watertight. We came to realize that many reputable experts across the world have significant misgivings about current theories concerning the Moon's origins that, as we have shown in this chapter, they were quite willing to voice publicly.

No matter how much the advocates of the Big Whack theory may claim they have solved the puzzle that is the Moon, it is quite obvious that this claim is far from being true. The Moon remains, to borrow the words of Winston Churchill, 'a riddle wrapped in a mystery inside an enigma'.

CHAPTER FOUR

Walking on the Moon

'We choose to go to the moon.'
President John F Kennedy: September 12th,
1962

After the end of the Second World War, rocket scientists from Germany were 'liberated' by both the United States and the Soviet Union, and by the beginning of the 1950s these experts were put to work on creating weapons of various sorts that would fuel the Cold War between the Eastern communists and the Western capitalists. On the American side the most famous of the German experts was Vernher Von Braun who had created the V1 and V2 rockets for Nazi Germany and who eventually went on to design the Saturn V rocket that would take people to the Moon.

At the outset the USA focused its attentions on developing new types of small but immensely powerful hydrogen bombs based on nuclear fusion whilst the USSR continued to refine the older and much heavier fission bomb. The Soviets therefore had to develop more powerful rockets and the R-7 missile, capable of carrying a five-tonne warhead, was the result. Their Chief Designer, Sergei Korolyov, realized that these rockets would also be capable of putting a one-and-a-half tonne satellite into Earth's orbit and he put forward his plan for such a mission.

Korolyov's project was well under way when news came that the US was developing its own satellite launch, known as Project Vanguard. This new challenge set up a 'race to space' and Korolyov's main satellite project was temporarily suspended as all efforts became focused on the early launch of a smaller artificial satellite that could be built far more quickly. Sputnik lifted into the skies on October 4th 1957.

This first spacecraft was a forty-pound sphere that carried a simple transmitter so that it could make meaningless, but technical sounding, bleeping sounds at which the world could marvel. The acclaim and sheer excitement caused by Sputnik's success led the Soviet leader, Nikita Khruschev, to demand more high-profile stunts rather than a return to serious science. The team responded immediately by screwing together the original Sputnik's backup spares to create a second Sputnik. They had only a few weeks as they were instructed that the next launch must happen before November 7th – the fortieth anniversary of the Great October Revolution.

Sputnik 2 was something of a botched job but it captured the imagination of the planet because it took off four days ahead of the anniversary and, amazingly, it was carrying a passenger: a dog called Laika. Unfortunately for this canine hero, her ticket was strictly one way because this hastily assembled craft had no mechanism for a controlled return to Earth – so the animal was destined to die in orbit from the outset. It is thought that she lived for four days in space before suffering a painful death as the cabin overheated. The fatality was part of the plan and the mission was considered a success as it proved that a living creature could survive the journey into orbit. So despite the fact that Sputnik 2 was initiated as a publicity stunt it was an important prelude to a human being making the trip.

The first two Sputniks were therefore politically inspired projects carried out by Sergei Korolyov under orders from the Kremlin and it was not until May 15th 1958 that his original spacecraft was launched – now designated Sputnik 3. This was a serious piece of equipment that was an automated scientific laboratory. It carried twelve instruments providing data on pressure and composition of the upper atmosphere; concentration of charged particles; photons in cosmic rays; heavy nuclei in cosmic rays; magnetic and electrostatic fields; and meteoric particles. And it was Sputnik 3 that first detected the presence of the outer radiation belts that surround the Earth.

The United States was highly embarrassed by the Soviet achievements, and particularly so because it was having little success with its own rocket launchers. So many of them blew up on the launch pad or during takeoff that the world's press variously dubbed the American space mission 'Kaputnik, Flopnik, and Stayputnik'.

In the summer of 1958 the Western world was rocking and rolling to Elvis Presley's 'Hound Dog', 'Heartbreak Hotel' and 'Jailhouse Rock' whilst the politicians of the ex-Russian territory of Alaska were lobbying to be accepted as the 49th State of the Union. In Washington, however, the US government's main focus was on something much more important – a new idea that was going to be a grand solution to a double-edged problem.

Their first concern was Sputnik. These high-profile launches had very effectively announced to the world that Soviet scientists were smarter than American ones and it was also implicit that the 'bad guys' had the technology to deliver heavy nuclear weapons around the planet. America had fallen well behind in the race for definitive military advantage and the idea of a 'first strike' by the Soviets suddenly seemed possible and, for some, even probable given the USA's current inability to respond in kind.

The second problem was one of internal power blocks. The US Army and Navy were politically untouchable and each had separate rocketry programmes causing duplication of effort that was dramatically slowing down the rate of overall progress. In the light of all this, Congress decided to side step military fiefdoms and set up a new organization to oversee and coordinate American space research.

Accordingly the National Aeronautics and Space Administration (NASA) was formed on October 1st 1958 and the idea of putting a man into space was immediately outlined, and given the title 'Project Mercury'. But it was a race they were destined to lose because on April 12th 1961 cosmonaut Yuri Gagarin became the first human to travel into space.

Gagarin's 108-minute voyage took him once around the planet, although he was not allowed to operate the controls because the effects of weightlessness had only been tested on dogs, and scientists were concerned that he may not be able to function properly. Consequently, ground crews controlled the mission with an override key provided just in case of an emergency.

NASA responded quickly by sending the astronaut Alan Shepherd on a ballistic trajectory sub-orbital flight to an altitude of 116 miles, returning to Earth at a landing point just 302 miles down the Atlantic Missile Range. America's first manned space flight was a fifteen-minute sky rocket event that was nowhere near the same league as Yuri Gagarin's 25,000 mile, high-speed voyage into Earth's orbit.

The race to get a man into space had been won by the USSR but there was a second, more ambitious competition running in parallel. Reaching for the Moon!

At first these were half-hearted attempts to get some metal, any bit of metal, onto the Moon. It had started with the first Pioneer rocket launched in 1958 by the United States – which lasted a full seventy-seven seconds before disintegrating into a giant fireball. A few months later the USSR launched Luna I, which performed beautifully but unfortunately missed the Moon and headed into solar orbit. In September 1959 the USSR managed to hit the bull's-eye when Luna 2 became the first craft to land on another celestial body, slamming into the Moon's surface just east of the Sea of Serenity. Before the impact Luna 2 was able to report back that there was something very odd about the Moon – it did not seem to have a magnetic field.

The next Soviet craft, Luna 3, made a great stride forward by swinging around the Moon, taking photographs of the 'dark' side before heading back to Earth in April 1960. The Americans meanwhile had failure after failure.

Nikita Khrushchev was pleased with the way that his nation was winning the space race and when Yuri Gagarin had orbited the Earth his propaganda machine went into overdrive to ensure that the world knew how superior his space engineers were. America's newly elected President was no slouch when it came to inspiring the public and John F Kennedy decided to take control of the situation by announcing that the real battle was to put men on the Moon. Despite a history of underperformance in space technology, he rather bravely publicly pledged to land a man on the Moon before the end of the 1960s.

Many American Ranger and Soviet Luna spacecraft headed for the Moon during the decade but a large number missed and others crashed onto the lunar surface either by accident or sometimes by design. But it was the USSR, once again, that made the next breakthrough when Luna 9 became the first spacecraft to make a controlled landing onto the surface of another celestial body on February 3rd 1966.

A significant part of the problem was the weird nature of the Moon's mass that was not at all what was expected. Instead of a generally constant gravitational field such as the Earth exhibits across its surface, the Moon is an inconsistent, lumpy ball that has huge variations in gravity from region to region.

As we have discussed, a pendulum swings with fairly regular precision on the Earth, with only quite small variations in swing rate because of the bulging of the planet at the equator. This is due to the fact that a person standing at sea level at the equator is a little further away from Earth's dense core than someone closer to one of the poles. Using a pendulum on the Moon would not produce any meaningful result because of what are known as 'mascons'.

The term mascon is an abbreviation for 'mass concentration' – regions of the Moon that have hugely dense material below the surface, rather than in the core as everyone would naturally expect. These mascons made it very difficult for spacecraft to orbit close to the Moon without continual adjustments to compensate for the variations in gravity. Some observers believe that it was this gravitational minefield that caused all of the problems for the early probes that were directed on the basis of a homogeneous gravity.

The existence of mascons was discovered after Lunar Orbiter 1 went into orbit around the Moon on August 14th 1966 and sent back high-quality images of over two million square miles of lunar surface, including the first detailed images of potential landing sites for the planned Apollo missions.

This new discovery of gravitational 'hotspots' on the Moon had an impact on a man who is arguably the greatest science fiction writer of all time and an acknowledged inspiration to NASA. Arthur C Clarke combined forces with film director Stanley Kubrick to write and shoot the most realistic space adventure ever. When their film *2001: A Space Odyssey* premiered in

April 1968, it stunned audiences across the world with its beautifully produced vision of the future.

The plot of the film starts millions of years ago when our ancestors were still apelike creatures without speech or tools. There is a visitation from some undisclosed power in the form of a jet-black and perfectly finished rectangular monolith that stands upright. When touched by the probing fingers of the gang of primates at dawn the monolith somehow remaps their brains to begin a process that will take these proto-humans on the evolutionary road to intellectual development. As the camera pans up the length of the monolith the Sun and the Moon appear directly overhead as though an eclipse is about to occur. The scene then leaps forward to the beginning of the twenty-first century when a powerful magnetic anomaly is discovered just below the surface of the Moon in the Tycho crater and excavations are carried out to discover what is causing the effect. A black monolith, some four metres tall is uncovered and a team of experts sets out from Earth to investigate the clearly artificial phenomenon.

The team travel to the Tycho crater as the Sun rises and wearing spacesuits they walk down a ramp into the pit where the monolith stands just a few metres below the surface. Like the man-apes millions of years earlier the team leader, Dr Floyd, is mesmerized by this alien structure and he touches it with his gloved hand. A moment later a ray of sunlight comes over the edge of the pit and strikes the monolith, signalling the end of the dark lunar night that lasts for two Earth weeks. This time, as we look up the monolith we see the Sun and Earth hovering directly above and almost touching. Then suddenly, the object transmits a signal in the direction of one of the moons of Jupiter (in Clarke's novel version this was changed to Iapetus, one of Saturn's moons).

The ingenious idea that Clarke put forward here was astonishingly close to the real-world discovery of the lunar mascons that had been made around the time he was writing. The similarity between Clarke's magnetic anomaly and the gravitational anomalies are obvious. We wonder whether Clarke was aware of the newly discovered mascons and whether that gave him the idea of a kind of trip switch placed on the Moon in the extreme past by some alien intelligence to trigger a signal that told them that creatures from the Earth had become smart enough to reach the Moon and spot a serious abnormality.

What a brilliant concept!

If an alien intelligence had indeed been responsible for the evolution of humans from ape to technologist, then what better way would there be of setting up an alarm system to confirm our intellectual 'arrival'.

At the time that Clarke and Kubrick's film was first capturing the imagination of a generation, no human had yet reached the Moon. But the following year, with less than six months to go to the late President Kennedy's deadline, Commander Neil Armstrong stepped out onto the surface of the Moon on July 20th 1969 with his famous but slightly misdelivered line:

> 'That's one small step for man, one giant leap for mankind.'

At this point we must mention that there are some people who seriously believe that NASA faked the Moon landings on a film set just like the one used by Stanley Kubrick. The evidence they produce looks reasonable at a casual glance; assuming you know nothing at all about photography or the facts relating to lunar conditions. These ideas suddenly leaped into the public imagination on February 15th 2001 when Fox television in the USA broadcast a programme called *Conspiracy Theory: Did We Land on the Moon?* The thrust of the show was that NASA technology in the 1960s was simply too primitive to have taken men to the Moon, and because they were so close to President Kennedy's politically important deadline they fabricated the entire mission in a movie studio.

To them the fraud was obvious. They point out that shots of the astronauts on the lunar surface show a completely black sky without any stars. Had this proved too difficult for the set constructors to fake they ask? The answer is actually very simple. As any proficient photographer knows, it is difficult to capture something extremely bright and something else extremely dim in the same shot. This means that for the stars to be visible, the lunar surface and the astronauts would have been burned out into a white blaze; the emulsion on a piece of film does not have enough dynamic range to capture both ends of the brightness scale simultaneously.

Amongst the other pieces of 'evidence' was the issue of the flapping flag. The NASA set designers were apparently so dumb that they allowed a stiff breeze to waft through the studio causing the flag that the astronauts planted to wave about. As the Moon has no atmosphere this is said to prove that it was filmed on Earth.

The fact is, the flag waved about so much precisely because there was no atmosphere. When astronauts planted the flagpole they rotated it back and forth to ensure that it penetrated the lunar surface causing the flag to wobble from side to side on its supporting frame. On Earth the presence of an atmosphere quickly dampens this motion as the surrounding air absorbs the energy from the moving flag, whereas in an airless environment the flag has nothing to dampen its motion. It could therefore keep going for many hours before the energy finally dissipated.

So anyone who has seriously looked into the case for and against the actuality of the Moon landings cannot fail to reject every one of the strands of evidence put forward by the conspiracy theorists. We do believe that conspiracies happen, because people will conspire together for all kinds of reasons – but the Apollo 11 mission was certainly not one of them.

We can be certain that twelve astronauts walked on the Moon between 1969 and 1972 and that they brought back 842 pounds of the Moon in the form of rocks, core samples, pebbles, sand and fine dust from six different exploration sites.

The last human being to walk on the Moon was Eugene Cernan in December 1972 and the information gathered over those three years, and later by Russian unmanned craft, has greatly increased our knowledge of the Moon. But it has also posed as many questions as it has answered.

It was expected that the samples of Moon rock would prove one of the existing theories about the Earth–Moon system. If the rock from the samples had been substantially different from rocks on Earth, then it was likely that the Moon had originated in some other part of the solar system and had been captured by the young Earth. If the Moon was identical in every way to the Earth, then it was likely they had both come into existence together and at the same time. However, it soon became apparent that both theories had to be wrong and no logical explanation for the Moon, being what it is and where it is, exists even now.

The convoluted 'Left hand/right hand double big whack' theory tends to crudely fill the void, to prevent us worrying too much about this hole in our knowledge of our planet and its neighbour. Whilst most people believe this rather unlikely hypothesis to be true, the people involved with developing it acknowledge that it is improbable. All existing theories of the Moon's origin have problems and the University of Wisconsin has pointed out that those for the Big Whack include:

- It requires that the entire Moon be initially molten and accreted from devolatilized material i.e. it does not account for the Moon's lower mantle's apparently largely undifferentiated composition.
- It requires that the impactor be accreted from the same oxygen reservoir as the Earth (a previous moon of Earth?).
- It does not account for a necessary density reversal below the upper mantle.
- It requires that differentiation of the Earth and the impactor, and their impact, occur within the 5HF/W 55-million-year model age for the lunar magma ocean.

- It does not account for the cumulative effect of many large impactors on the Moon's non-parallel rotational axis.
- It does not account for the necessary chronology of tidal separation of the Earth and moon origin of the Moon.

There is also another major problem with this scenario revealed by the issue of the ongoing slowing down of Earth. Very precise astronomical measurements, some of them dating back to the observation of eclipses 2,500 years ago, indicate that the day is increasing in length by about one or two thousandths of a second per day per century. It has been thought that this tiny lengthening of the day was entirely due to the friction of the tides caused by the Sun and the Moon. But when attempts were made to predict changes in the apparent position of the Moon on the basis of this effect alone, it was found that the calculations did not agree with the observations at all. Another factor must be at work as well.

That factor was that iron is sinking to the core of the Earth, changing the moment of inertia and thereby the length of the day. When this was taken into consideration and calculations were made on the basis of both the tides and the changing moment of inertia due to sinking iron, the sums did agree with the observations. But in order to make the calculations agree, it was necessary to postulate a flow of 50,000 tonnes of iron from the mantle to the core of the earth every second!

Staggering though this volume of flow is, it would still take 500 million years to form the metallic core of the Earth and some calculations indicate that it may have taken as long as two billion years. If this reasoning is correct, which it appears to be, the Earth was made initially with large amounts of iron in its exterior parts. As the Moon was formed at a very early stage in the Earth's existence (and possibly before), any material knocked off the surface by a major impact would contain large amounts of iron – which it does not.

The Big Whack theories are simply the best of all the impossible explanations for the existence of the Moon.

It is widely accepted that despite the intense investigation that has gone into understanding the Moon, and for all we know about its surface and the composition of its rocks, we are as much in the dark concerning its origins as we were before the first projectile left the Earth's atmosphere.

As we have discussed, the oxygen isotope investigation proved that both Moon rocks and Earth rocks must have developed at exactly the same distance from the Sun, so the Moon definitely wasn't a captured asteroid. The Moon has its fair share of the elements found on Earth but not in the same proportion. The Moon is substantially lacking in heavy metals when compared with the Earth, which accounts for its large size but small mass.

But it was the Apollo missions that identified something else that was weird about the Moon.

'Houston, we've got a problem'

The first two Apollo crews had landed out on the smooth lunar mare, the lava seas that are relatively young by lunar standards, and now NASA wanted to visit a site where they could study the older parts of the Moon, which meant the rugged highlands. Although NASA was not ready to commit a lunar module (LM) to a landing in highly rocky terrain, the site selection committee was very interested in a place called the Fra Mauro Hills in the middle of the Ocean of Storms, which seemed like a fairly smooth section of the highlands.

Commander Jim Lovell along with Jack Swigert and Fred Haise were chosen for the Fra Mauro mission as the crew of Apollo 13. The launch, on April 11th 1970, went well, allaying the worst fears of those who were concerned about a mission with the unlucky number thirteen.

Then, fifty-five hours and fifty-five minutes into the mission (and on the thirteenth day of the month) all three astronauts heard and felt what they described as a 'pretty large bang' on board the spacecraft. The crew and the ground controllers made a rapid assessment of the health of the spacecraft and it was obvious that two of the three fuel cells in the service module were dead. No one knew exactly what had gone wrong but there was no doubt that the crew were in serious danger.

To survive they needed enough power, oxygen, and water for a four-day trip around the Moon and back to Earth, and it now looked as if these commodities were going to be in very short supply. Oxygen and hydrogen were normally combined in the fuel cells to produce electricity and water and both oxygen tanks were rapidly losing pressure so even the remaining fuel cell wouldn't last long. In addition to short supplies of these basic commodities, without power in the command module, they would have to rely on the LM environmental control system to remove excess carbon dioxide from the cabin. And to add to their many woes, the main engine now had no power supply.

However, the flight crew and ground personnel all realized just how lucky they had been. As desperate as the situation was, the accident had come early in the mission and they still had their fully stocked lunar module as a resource. The LM had an engine that could be used to put the crew back on a homeward path, and it carried just enough water, oxygen, and power for the four days they need to fly around the Moon and head home.

As the stricken spacecraft swung behind the Moon, 164 miles above the surface, contact with the Earth was lost until it emerged on the other side and was again picked up by tracking stations. The following words were heard: 'The view out there is fantastic ... You can see where we're zooming off.'

At 8:09pm EST on April 14th, Apollo 13 turned for home and the third stage of the Saturn V launch vehicle, weighing fifteen tonnes, was sent crashing into the Moon. As planned it struck the Moon with a force equivalent to 11 1/2 tonnes of TNT. The impact point was eighty-five miles west-northwest of the site where the Apollo 12 astronauts had set up a seismometer.

NASA reports demonstrate the reaction of scientists on Earth as the Saturn V hit the lunar surface – 'The Moon rang like a bell.'

In November 1969 the Apollo 12 crew had sent their lunar module crashing into the Moon following their return to the command craft after their lunar landing mission. That lunar module had struck with a force of one tonne of TNT causing the shock waves to build up to a peak in eight minutes and then continue for nearly an hour. The seismic signals produced by the impact from Apollo 13 were twenty to thirty times greater and lasted four times longer than those resulting from the earlier LM crash. This time, peak intensity occurred after seven minutes and the reverberations lasted for three hours and twenty minutes, travelling to a depth of twenty-five miles, leading to the conclusion that the Moon has an unusually light core or possibly no core at all.

At the time Houston remarked to the Apollo 13 crew: 'By the way, Aquarius, we see the results now from 12's seismometer. Looks like your booster just hit the Moon, and it's rocking a little bit.'

NASA reports how the information from these two artificial moonquakes led to a reconsideration of theories proposed about the lunar interior. Among the puzzling features, they say, are the rapid build-up to the peak and the prolonged reverberations, because nothing comparable happens when objects strike Earth.

When Chris was in Seattle a few years ago he had a meeting with Ken Johnston who had worked for Brown-Root and Northrop, which was a consortium between the Brown-Root Corporation and the Northrop Corporation at the Lunar Receiving Laboratory. The company was one of the prime contractors for NASA at the time of the Apollo missions and Ken was supervisor of the data and photo control department. Ken told Chris that at the time of the impact created by the Apollo 13 launch vehicle the scientists were not only saying that 'the Moon rang like a bell', they also described how the whole structure of the Moon 'wobbled' in a precise way, 'almost as though it had gigantic hydraulic damper struts inside it.'

This ringing effect caused many people to pick up on speculation that had been going on for years that the Earth's Moon could be hollow. Back in 1962 Dr Gordon McDonald, a leading scientist at NASA, published a report in the Astronautics Magazine where he stated that analysis of the Moon's motion indicated that the Moon is hollow.

Dr Sean C Solomon, who was Professor of Geophysics at MIT and is the Director of the Terrestrial Magnetism Department, Carnegie Institution of Washington as well as the Principal Investigator for Carnegie's research as part of the NASA Astrobiology Institute, has said: 'The lunar orbiter experiments vastly improved our knowledge of the moon's gravitational field ... indicating the frightening possibility that the moon may be hollow.'

Why should this be frightening?

Carl Sagan, Professor of Astronomy and Space Sciences and director of the Laboratory for Planetary Studies at Cornell University hinted at the answer when he said, whilst discussing the moons of Mars, that 'It is well understood that a natural satellite cannot be a hollow object.'[14]

The problem therefore is simple – if the Moon is hollow, someone or something manufactured it.

But the debate continues. A team from the University of Arizona in Tucson has detailed the results of their interpretation of data from the

Lunar Prospector magnetometer where they estimate that the moon does have a tiny metal core that is roughly 420 miles (680km) across, plus or minus 112 miles (180km). Their team leader, was Lon Hood. 'We knew that the Moon's core was small, but we didn't know it was this small,' Hood said. 'This really does add weight to the idea that the Moon's origin is unique, unlike any other terrestrial body – Earth, Venus, Mars or Mercury.'[15]

So, it is possible that the Moon is hollow at its centre or has a very small core. There is also the possibility that it has voids in its make-up just as it has the super-dense zones we call mascons. But it seems that the structure is unusual whatever the case turns out to be.

The main argument against the idea of a hollow Moon that we found repeated time and again, was that there was no theory of the Moon's origin that could explain such a circumstance. The argument goes: 'Because we can't explain how a natural satellite can form with a hollow centre – it cannot have one QED.'

This standpoint is fair enough – if you accept its founding premise, that the Moon is natural. And who would not make such an assumption?

But as we put aside all of our preconceptions about what can and cannot be, we have to accept that solid objects do not ring like a bell – but hollow ones do.

Hollow or not, we decided to look more closely at the mechanics of the Moon.

CHAPTER FIVE

The Bringer of Life

We had seen just how peculiar the Moon is, in so many ways. Our next step was to look into how our next-door neighbour in the cosmos actually affects life on Earth.

First of all we could not ignore the myth that the full Moon brings out madness and other evils in the form of more violence, more suicides, more accidents and more aggression – ideas that are possibly as old as history itself. The belief that the full Moon causes mental disorders and strange behaviour was particularly widespread throughout Europe in the Middle Ages.

But is there any scientific evidence to support these beliefs?

There have been many investigations into the subject and some have produced surprising results. Research carried out by a medical team at a hospital in Bradford, England, set out to test the hypothesis that the incidence of animal bites increases at the time of a full Moon. Using retrospective observational analysis at their accident and emergency department they investigated the pattern of patients who attended from 1997 to 1999 after being bitten by an animal.

The number of bites in each day was compared with the lunar phase in each month and they found that the incidence of animal bites rose significantly at the time of a full Moon. With the period of the full Moon as the reference point, the incidence rate ratio of the bites for all other periods of the lunar cycle was significantly lower. They concluded that the full Moon is associated with a significant increase in animal bites to humans.[16]

Of course, we must remember that correlation does not equate to causation. The pattern they found may be a strange statistical blip or, even if it is real, it could be entirely coincidental with the phases of the Moon. Without any suggestion of how the Moon could cause an increase in animal aggression towards humans, it is not possible to consider any connection as proven.

Another study looked into human aggression and the lunar synodic cycle occurring in Dade County, Florida. Data on five aggressive and/or violent human behaviours were examined to determine whether a relationship existed between the two. These included looking at the pattern of homicides, suicides, fatal traffic accidents, aggravated assaults and psychiatric emergency room visits.

The team concluded that homicides and aggravated assaults demonstrated a statistically significant clustering of cases around the full Moon. Psychiatric emergency room visits clustered around the first quarter and showed a significantly decreased frequency around new and full Moon. The suicide curve showed correlations with both aggravated assaults and fatal traffic accidents suggesting, they say, a self-destructive component for each of these behaviours. The existence of a biological rhythm of human aggression, which resonates with the lunar synodic cycle was postulated.[17]

Whilst these investigations were carried out carefully and scientifically it is important to remember that there are dozens of other studies that have failed to identify similar correlations. If there is some substance behind lunar myth it is yet to be proven. However, we feel that such a relationship is not beyond reason as the Moon exerts considerable gravitational effects on the Earth creating the tidal movements of the waters of our oceans, and humans are made up of nearly eighty per cent water. Whether or not lunar cycles affect our lives; solar ones certainly do.

The Four Seasons

At the time of writing these words the leaves on the trees here in Britain are beginning to be tinged with brown. The days are growing shorter and the nights are getting longer. As this happens, the average temperature each day begins to fall and much of our flora and fauna goes into a dormant state.

Of course, the same seasonal change is happening all across the northern hemisphere at latitudes between the Tropic of Cancer and the Arctic Circle. Meanwhile, countries in the southern hemisphere are entering spring and new growth is beginning to stir as the days lengthen and the average daily temperature increases. All of us who do not live on or near to the equator are familiar with the pattern of the changing seasons and the effect that these cycles have on the way we live our lives. To our ancestors in northern parts of Europe, Asia and America, the onset of winter must have been a time of fear and doubt, whilst the first buds of spring would have been a merciful relief with the signal that there would soon be fresh food to eat.

What most of us don't stop to think about is why seasons happen at all. It is a common misunderstanding to imagine it has something to do with how close the Earth is to the Sun. It is not – it is due to the angle of the planet in relation to the Sun, which is about 22.5 degrees from what might be described as a vertical position. The diagram below shows how the Earth would look if it was standing upright as it goes around the Sun, which would mean that the equator of the Earth would always point straight at the equator of the Sun.

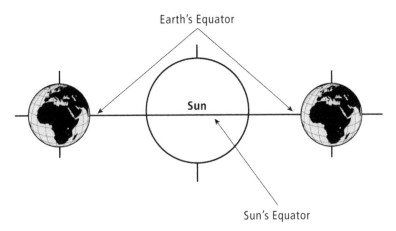

Figure 5

If our planet really did stand in this position, the bulge of the Sun's equator and that of the Earth would be closer together than the Sun's poles and the Earth's poles. The result of this would be a superhot equatorial temperature on the Earth, whilst the polar regions of the Earth would be much colder than they presently are. Strangely enough it's not so much a case of the difference in distance between the Earth and the Sun that matters; it is more to do with the thickness of the atmosphere above any given part of the Earth in relationship to the direction of the Sun. In the imaginary situation above, sunlight has to travel through far more atmosphere to get to the poles of the Earth than it does to reach the equator, thus greatly reducing the heat.

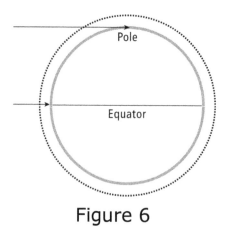

Figure 6

Another important factor that reduces the heat at the poles is diminished power density, where the Sun's energy is dissipated across a greater area as the Earth curves away from an upright position. For example, a circle of sunlight with a one-kilometre diameter will hit the Earth's surface as a near perfect circle at the equator, but in extreme northern or southern latitudes it will be distorted into a long oval due to the curvature of the planet. This means that the heat of the sunlight at the poles will be spread over several times the area and therefore be several times weaker.

The planet Mercury is an excellent example of a world that is standing virtually upright, in relation to its orbit around the Sun. Apart from the fact that little Mercury is so close to the Sun, its angle of inclination, or 'obliquity' as it is more properly called, would make it a very uncomfortable place for humans. If it were possible to stand on Mercury during one of its very short eighty-eight-day years, the Sun would rise due east every day (which is equal to fifty-eight Earth days) at the equator and set due west. Mercury has equatorial temperatures that would keep lead boiling, yet probes sent from Earth have shown that the polar regions of Mercury are constantly covered in ice.

So, if the Earth were in this upright mode, life would be almost impossible across much of the planet, with extremes of temperature providing only a narrow band suitable for mammals such as humans to survive. Even then, the sea and air currents would move wildly between the hot and cold zones causing catastrophic weather conditions with regions of permanent rainfall and others with none at all. Hurricanes and tornadoes would ravage many areas and overall it seems extremely unlikely that higher life forms would ever develop on such a planet.

Now consider another imaginary scenario in which the Earth is tilted on its axis a full 90 degrees relative to its orbit around the Sun so that one pole faces the Sun at all times.

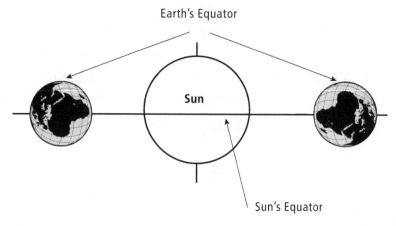

Earth's Equator

Sun

Sun's Equator

Figure 7

One of the poles, say the South Pole, would be permanently in daylight – stuck for ever in a position equivalent to noon on midsummer's day in central Africa. The Sun would blaze down from directly overhead every minute of every day! The North Pole on the other hand, would be in a state of constant midnight. Indeed, all of the northern hemisphere would be in constant night and the southern in constant day.

The dark side of the planet would never warm up and it would be frozen solid with temperatures far below anything we actually experience. The region that is currently between our equator and the Tropic of Capricorn would see the Sun circling right around, low on the horizon once each day. Because of the angle of the sunlight through the atmosphere, there would be very little warmth getting through and the entire region would be covered in glaciers and swept with snowstorms driving down from the dark northern hemisphere.

Antarctica would be utterly uninhabitable, being far hotter than anywhere on our planet as we know it today. Only the southern tip of South America, Tasmania, New Zealand and maybe the southern section of Australia would have temperatures that were within a tolerable range. But it is hard to imagine what kinds of terrible weather anyone living there would have to endure, with freezing ocean currents moving from the north and very hot ones arriving from the south. A state of permanent fog seems certain; which would in turn block out the Sun.

If the Earth orbited the Sun in either of the two modes we have just described, there would be no seasons at all – and almost certainly no higher life forms.

Thankfully we do have seasons, courtesy of the fact that the Earth is actually at an angle of around 22.5 degrees relative to the equator of the Sun. And that angle is maintained by the Moon, which acts as a gigantic planetary stabilizer.

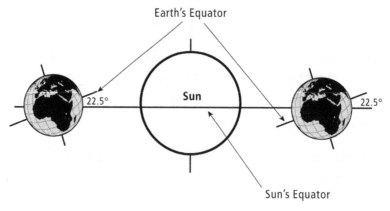

Figure 8

Because of this tilt, the northern hemisphere experiences summer when the Earth is on that part of its orbit that angles it more towards, the Sun. Therefore the Sun rises higher in the sky and is above the horizon longer, and the rays of the Sun strike the ground more directly. Conversely, when the northern hemisphere is oriented away from the Sun, the Sun only rises low in the sky, is above the horizon for a shorter period, and the rays of the Sun strike the ground more obliquely.

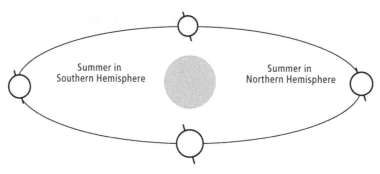

Figure 9

Whilst it is true that the extreme polar regions of the Earth are frozen throughout the year, the tilt angle of 22.5 degrees ensures that most parts of the Earth's surface get a fair share of warmth throughout each year. This in turn means that by far the vast majority of water on the surface of the planet remains in a liquid state. All of life is utterly dependent on water and cannot exist without it. The band of temperatures at which water is liquid is really very narrow. The oceans of the Earth would freeze at around 1.91°C, with boiling point occurring at 100°C.

The Earth is therefore extremely well balanced. The coldest temperature ever recorded was -89.2°C (-128.6°F) at the Vostok Station in Antarctica and the highest was 58°C (136°F) at El Azizia in Libya. That is a range of absolute extremes of less than 148°C, which is very little indeed in terms of the entire spectrum. The coldest anything can get is known as 'absolute zero' when all molecular motion stops. This occurs at a rather chilly -273.15°C (-459.67°F).

On the other hand there is no known upper limit for temperature but the hottest temperature in our solar system is the Sun's core, which comes in at an impressive 15,000,000°C (27,000,000°F).

The normal temperature range on Earth is such that there are very few parts of the globe that cannot support human life. We have a normal range of body temperature between

36.1 to 37.8°C (97 to 100°F) and yet the Inuit people live happily within the Arctic Circle and the Bedouin travel the deserts of North Africa.

The world's average temperature fluctuates slightly around the 14.5°C (58°F) mark, which is comfortable for physical work. Of course, some people will say that the world 'is' that temperature and that we would not have evolved as we have if it were any different – but this is flawed logic. We could just as well have evolved in a world where only small sections of the planet were available to us to inhabit. No other known planet has such a narrow temperature band – and a range of temperature that permits water to be liquid most of the time.

In fact water is a very curious substance altogether. On Earth we can see it at the same time in its three states – as solid ice, as liquid water and as a gas in clouds. Each water molecule is composed of just two atoms of hydrogen and one of oxygen and yet it acts as a universal solvent with a high surface tension.

Perhaps most surprising of all is how its density changes. Water has its maximum density at 4°C which means that it not only gets lighter as it warms from that point – it also gets lighter as it cools. As everyone knows, warm water rises as convection currents but it is also true that ice floats. Other planets in our solar system may have ice or steam but only the Earth is awash with life-giving liquid water.

Liquid water has been absolutely crucial in creating the world we know today and, as far as is known, life cannot exist without it. As surely as plate tectonics and the Earth's hot core constantly create new mountain ranges, via volcanoes and the pushing up of mountains as land masses meet, so water is mainly responsible for flattening them again. Constant weathering crumbles away the rocks as mountains age and water, in the form of rain, ice and snow, is primarily responsible. Liquid water, as streams and rivers, also disperses the weathered rock, carrying it down to the plains where it is distributed across flatter land, bringing much needed nutrients to nourish life. Even more nutrients are carried by the rivers to the oceans where they offer the necessary food for aquatic plants that stand at the bottom of the oceanic food chain.

Of course, none of this would be possible if the vast majority of water on the Earth was not in a liquid state. Only two per cent of Earth's water is locked up in glaciers and the icecaps, with ninety-seven per cent being the water of our seas and oceans and just one per cent available for human consumption as fresh water. With only a small change in the overall temperature of the Earth, or an alteration in the seasonal patterns, the nature of the water on our planet would change. As we have seen, a more pronounced planetary tilt could well lead to a freezing of the oceans. This would result in an overall loss of temperature at the surface of the planet, with even greater freezing.

On the other hand, if the Earth were not tilted at all, the equatorial regions would become unbearably hot and weather patterns across the planet would be radically changed. In addition, the biodiversity, that scientists are now certain has been so important to our evolution, might never have developed in a world with more polarized areas of temperature.

It has therefore been vital for our existence that the tilt of the Earth has been maintained at around 22.5 degrees for an extremely long period of time, and yet, bearing in mind the composition of the planet this is a very unlikely state of affairs. Venus is the nearest planet to Earth and the most similar to our own, but it has toppled over in the past and other planets in the solar system show signs of having varied markedly in their tilt angle across time. The Earth is very active internally and highly unstable, yet, despite a few periodic wobbles, it keeps the same angle relative to the Sun.

Astronomer Jacques Laskar, a Director of Research at the National Scientific Research Centre (CNRS) and head of a team at the Observatory of Paris is in no doubt that the Earth would indeed topple over, if it were not for the presence of the Moon![18]

With computer modelling, Laskar showed in 1993 that all the other Earth-like planets (Mercury, Venus and Mars) have highly unstable obliquity, which, in the case of Mars for example, varies wildly across time between 0 degrees and 60 degrees. The same computer modelling indicates that in the case of the Earth the obliquity would vary even more, between 0 degrees and 85 degrees – but for the stabilizing influence of our incredibly large Moon.

Nobody knows for certain how long it would take for the Earth's obliquity to change significantly if the Moon was not exerting such a massive influence. There is a constant transfer of energy taking place between the two bodies, which in addition to stabilizing Earth's obliquity has also significantly slowed our planet's rate of spin. This constant obliquity has made the Earth a perfect crucible for advanced life by providing many millions of years of stability for life to develop from its simplest form to the complex patterns it adopts today.

Although the Earth is significantly more massive than the Moon, the Moon is still a very large body. Tides in Earth's oceans, seas and lakes are caused by the gravitational interaction of the Earth, the Moon and the Sun. Tides have an effect on dry land as well as oceans but this effect can only be detected by careful measurement. Solar tides (the point of greatest gravitational pull by the Sun) are twelve hours apart but since the Moon is also moving, lunar tides are slightly more irregular, occurring every 12.42 hours on average.

The height of tides in any particular part of the ocean is dependent on a number of factors such as the shape of any nearby landmasses and the depth of the seabed. In some areas of the world, tides hardly seem to lift the level of water at all – this is just as well for some low-lying places such as the islands of the Maldives in the Indian Ocean because their average height above sea level is less than one metre. In other places, like the British coast, tides can have a huge range between high and low water.

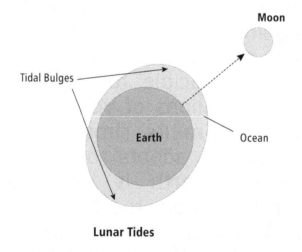

Figure 10. The Moon has sufficient gravity to pull a bulge of water from the oceans of the Earth closest to its position towards it. It also distorts the Earth, creating a corresponding bulge in the oceans on the opposite side of the Earth. Because of the Earth's rotation the bulge on the Moon side runs slightly ahead of the Moon.

Tides would not cease if the Moon were not present because they are also created by the

Sun. However, they would be very much lower than they are now because although the Sun is massive and the Moon much smaller, the Moon is extremely close and the Sun much more distant. It is the interaction of solar and lunar tides that makes it rather complicated to predict when tides will occur and how high or low they are likely to be.

The highest of the lunar tides occur when the Moon is either in its full or new mode, because at such times it is in line with the Sun and its gravitational forces are added to those of the Sun. Much lower tides are on the first and last quarters of the Moon when the gravity of the Moon and the Sun are working against each other.

Life in the tidal margins of the oceans and seas has evolved to take advantage of tides, either in a daily or a monthly sense. Some species of crabs for example, lay their eggs in the sand at the high-water mark at the time of the full or new Moon so that they will be safe from marine predators during incubation. There are also many creatures that leave the ocean on the high tide at night to scavenge in the intertidal margins, before seeking safety with the next high tide.

Many shellfish are absolutely dependent on the ebb and flow of the tides for the purpose of feeding and it was shown in the 1960s that oysters are sensitive enough to be aware of the Moon's position, either overhead or at the opposite side of the planet. Oysters, which obviously have no eyes, were taken from the ocean and placed in tanks in the Rocky Mountains where they began to open and close, as they would have done in the ocean, had it extended so far inland. Because other stimulus such as current or wave motion were absent, it suggests that they are able to feel minute increases and decreases in the gravitational pull of the Moon and the Sun.

If molluscs, our very distant evolutionary cousins, can somehow sense such astronomical movements – then there would seem to be no reason why humans would not be able to do the same. This just might point the way forward in investigating a possible causation for variations in human behaviour according to the phase of the Moon.

It probably is not too surprising that some creatures have learned to exploit tides, which are tiny these days in comparison with the remote past when the Moon was much closer to the Earth. The tremendous forces created by a very close Moon would have generated much heat and might even have caused parts of the Earth's surface to melt. However, this phase did not last all that long because the very transfer of energy that promotes tides is also causing the Moon to drift further and further away from the Earth. This happens because the Earth rotates around its own axis more quickly than the Moon revolves around the Earth. The rapid rotation means that the tidal bulge of the Earth forward of the Moon, (see figure 11) is always ahead of the Moon's position. The tidal bulge exerts a pull on the Moon and this increases the Moon's overall energy. Meanwhile, friction between the Earth's surface and its own oceans is actually slowing the rate of Earth rotation. It is not much, but it does amount to around 0.002 seconds in a century.

The end result of this dance will be that the Moon will continue to move away from the Earth until a situation of equilibrium is achieved, which is expected to happen in about fifteen billion years. The Moon will then be 1.6 times further out from the Earth than it is now and the Earth will have a solar day that is equal to the orbit of the Moon, which by then will be fifty-five days. However, we do not have to lose too much sleep about this eventuality because the Sun will have become a red giant about a billion years before that, at which time the Earth will have ceased to exist in any case.

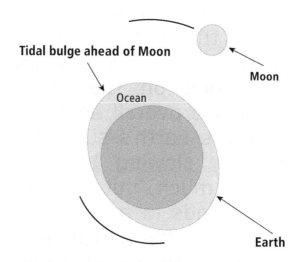

Figure 11. As the Earth revolves, it takes the tidal bulges with it, but because of the gravity of the Moon, the water in the bulges is trying to travel in the opposite direction. As a result, waves ground on the bottom of the oceans and on seashores, causing friction. The friction slows the Earth and the energy is passed to the Moon, which responds by speeding up. As it does so, the laws of physics dictate that its orbit must widen.

Over huge periods of time the relationship between the Earth and the Moon changes, so we find ourselves living in what amounts to a 'tiny snapshot' of the overall situation. At present the Moon takes 27.322 days to go around the Earth and because the Earth is also going around the Sun, full and new Moons are ruled by a slightly longer cycle that takes 29.53 days. Both these figures have been significantly different in the past and will be different again in the future but the changes are very slow and, according to NASA, the Moon is becoming more distant from the Earth by around 3.8cm per year.

Perhaps it is just as well that an expanding Sun will overtake us before the Moon does get to its final position relative to the Earth. By the time the Moon and the Earth reach their ultimate stations, the Moon will be too distant to exert enough influence on our planet to keep its obliquity steady. Bearing in mind the Earth's unstable core, this would almost certainly mean rapid and perhaps catastrophic changes in both obliquity and climate.

Neil F Comins, Professor of Physics and Astronomy at the University of Maine, has written about the consequences if the Moon did not exist. He explains that the Earth would be turning so fast that a day would take just eight hours and complex life would not exist yet. If higher life forms did eventually manage to evolve, such creatures would be very different to us without, for example, any communication through speech.[19]

One thing is certain then: no Moon would mean no humans!

CHAPTER SIX

The Living Earth

Humans are incredibly robust creatures considering we are little more than animated bags of water hanging on a mineral frame. We can withstand difficult conditions and even survive without food for many weeks, yet we die quickly without air to breath or with direct exposure to unusually high or low temperatures. It is thanks to eons of Darwinian evolution that we are perfectly designed for our environment – but perhaps we should not be too casual about the extraordinary good fortune that brought us to this point.

Every human is very special. We differ from other creatures, so we are told, because we are able to define ourselves by our own self-awareness resulting in a situation where there is a simple polarity to the Universe. We all know that: 'There is *me* and then there is everything else.' Each and every one of us is an emotional-intellectual island connected to that 'everything else' by the complex interaction of our five senses.

Two small regions of our skin have developed the ability to decode energy reflections in the form of sight, two more make sense of a cacophony of colliding compression waves in the gases around us giving us hearing. Then we have skin sensitive enough to tell us about shape and texture, a mouth that accurately differentiates between different chemical substances we are about to consume in the form of taste and we have an air inlet that can pick out the presence of a specific molecule within a million others in the atmosphere as the sense we call smell.

These five connection modes cause us to have interaction with the 'everything else' – especially other humans, so we do not exist alone. These points of stimulus combine to give life to the most remarkable array of aspects of self. Love, fear, loathing, compassion, laughter and countless other emotions make us special and mark us out as entities that are utterly different to the rest of creation.

But how and why have we become so spectacularly differentiated from other combinations of recycled stardust? What makes Neil Armstrong more special than the 3.5-billion-year-old rock he first lifted from the lunar surface?

Those with religious faith turn to their interpretation of God to explain the unexplainable and the more scientific amongst us turns to the Anthropic Principle. The good old 'Anthropic Principle' is less there to help us answer the BIG question than to avoid having to deal with it. It accepts the vanishingly tiny probability of human existence by stating that the rules of the Universe that produced us have to be exactly as they are or we would not be here to perceive them.

To us, this is rather like defining moving, emotionally stimulating music by merely expressing it as 'music that is good'. The statement is correct but it does not compare with the experience!

What the Anthropic Principle does is to stop us worrying too much about the fact that we really have no right to exist. Of the two approaches, anthropic or divine, at least the God scenario is an attempt to move the problem on a notch rather than utilizing a principle that seems to have been conceived to ignore it.

Most scientifically minded people probably subscribe to the theory that humans, like everything else, are the product of billions of years of random chance. However, the most famous scientist of all time, Albert Einstein, was very unhappy about nature being based on randomness. He said about quantum physics: 'God does not play dice.'

The more we looked into how our planet developed into a paradise for living creatures the more surprised we became. The miracle of life on Earth is due to our narrow temperature band that provides us with liquid water and, as we have explained, it is the Moon that is responsible for maintaining the perfect tilt that provides our benign climate. But amazingly, it was the very act of the Moon's creation that produced the first link in the chain of events that would lead the Universe to make you!

In 1911 a brilliant young scientist by the name of Alfred Lothar Wegener was browsing through the library of his university in Marburg, Germany, when he came across a scientific paper that listed a host of identical plant and animal species that could be found on opposite sides of the Atlantic. Although having obtained a PhD in astronomy at a very early age, Wegener was particularly interested in geophysics, a field of study that was in its infancy at the time.

Something in the paper caught Wegener's imagination and he began to spend time looking for other examples of similar plants and creatures separated by oceans. There was, at the time, no reasonable explanation as to how such a state of affairs could have come about. It had been postulated that the solution to this puzzle had to be land bridges that must have existed in very ancient times and that had allowed both plants and animals to move between continents. However, there were many examples that could not be explained in this way.

Wegener had also noted, as had others before him, how many cases there were in which the coastline of one continent looked as though it could fit snugly into that of another, such as the west coast of Africa and the east coast of South America. He also found that if the continental shelf is studied, rather than the apparent coastline shaped by current sea level, the fit is often very much better.

Alfred Wegener began to ask himself if the answer to these anomalies might lie not in land bridges but in the fact that the continents were once joined together in one large continent, and that this had somehow broken up and drifted apart. Later in his life he wrote about this process of logical deduction. 'A conviction of the fundamental soundness of the idea took root in my mind.'

Wegener spent a considerable period collecting further examples of extended flora and fauna and the available evidence continued to support his early theory. For example, he found the fossils of plants and creatures in places where the climate must have been significantly different when they were alive and flourishing, such as fossilized cycads – ancient tropical plants found as far away from the tropics as Spitsbergen in the Arctic.

From the weight of evidence he had collected, Wegener deduced that all the continents had once been part of a single landmass, which he chose to call 'Pangaea' – a Greek word meaning 'all the Earth'. He suggested that this super-continent had broken up and had begun to drift apart 300 million years ago. He called the process 'Continental

Drift' and although he wasn't the first to suggest that there had originally been a single continent, he was able to provide substantial evidence to back up the claim. Wegener first published his findings and his hypothesis in his book *The Origin of Continents and Oceans.*[20] Although it was brilliantly argued, his ideas were not widely accepted at the time.

A flood of scientific indignation broke over Alfred Wegener. This happened for a couple of reasons: firstly, his theory was revolutionary, which inevitably clashed with the conservative tendencies of other experts; and in addition, although Wegener was certain that continental drift must have taken place, he had no theory as to how or why this might have happened. The best he could suggest was that the continents, influenced by centrifugal and tidal forces as the Earth spun on its axis, were simply ploughing their way across the surface of the planet.

Dissenters pointed out that, if this was the case, the coastlines of the continents could hardly be expected to have remained so similar to the original 'fit' that it could still be observed. On the contrary, they would have been distorted beyond recognition. It was also suggested that tidal and centrifugal forces would be far too weak to move entire continents.

Poor Alfred Wegener didn't have the chance to look too much further into the matter; he died in 1930 whilst taking part in a rescue mission to deliver food to a party of explorers and scientists trapped in Greenland.

Wegener did have some notable supporters but in general his ideas remained on the shelf until as recently as the 1950s, by which time greater exploration and understanding of the Earth's geophysical makeup had begun to catch up with the idea of continental drift. The truth of the matter is that Wegener was wrong in terms of his suggested mechanism, but quite correct in his basic assumption. Rather than ploughing their way across the planet's surface, the continents 'float' on what is known as the 'asthenosphere', the underlying rock of our planet. This is under so much pressure and becomes so incredibly hot that it acts more like thick treacle than solid rock.

The Composition of the Earth

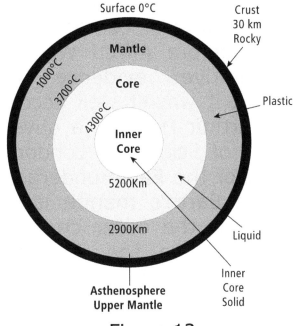

Figure 12

One of the factors that made Wegener's ideas more acceptable was the study of mountain ranges. An earlier position held by many experts had been the 'contraction theory'. This suggested that the Earth had begun its life as a molten ball and that as it cooled it had cracked and folded up on itself. This folding, the theory suggested, was what had created mountain ranges. The real problem with the contraction theory was that all mountain ranges should therefore be of the same age and it was rapidly becoming apparent that this could not be the case. Wegener had suggested that mountains were constantly being created as landmasses came into contact, exerting unbelievable pressure and pushing up land at or close to the points of contact.

Just a year before Alfred Wegener's death some corroborative evidence had been forthcoming, but it wasn't well accepted at the time. In 1929 Arthur Holmes, a physicist at the Imperial College of Science in London suggested that the mantle of the Earth undergoes 'thermal convection'. The Earth's mantle is that region immediately below the outer crust. It extends all the way down to the Earth's core. Its composition varies with increased pressure and temperature but it makes up the biggest part of the Earth.

Holmes knew that when a substance is heated, its density decreases. In the case of the mantle this would cause material to rise to the surface where it would gradually cool, become denser and then sink again. A similar process takes place with porridge that is boiling in a saucepan. Holmes was quite taken with Wegener's idea of continental drift and suggested that the tremendous pressures caused by thermal convection could act like a conveyor belt. This might cause the continents to break apart and to be 'carried' across the surface of the planet.

For years these ideas were dismissed, until knowledge caught up with the theories. By the 1960s there was a greater understanding of the 'oceanic ridges'–regions where, it was being realized, Holmes' thermal convection might actually be taking place. It was also realized that oceanic trenches occurred, together with arcs of islands, close to the continental margins. All of this meant that convection was not only probable but certain. Two other scientists, R Deitz in 1961 and Harry Hess in 1962 separately published similar hypotheses based on mantle convection currents, and continental drift became universally accepted.

Deitz and Hess between them modified Holmes' original theory of convection and came eventually to their own mechanism for continental drift, which is based on what they termed 'seafloor spreading'. This spreading, it is suggested, begins in the mid-oceanic ridges. These are huge mountain ranges in the middle of the Earth's largest oceans. So large are the mid-oceanic ridges that they are higher than the Himalayas and are more than 2,000 kilometres wide. Associated with the ridges are great trenches that bisect the length of the ridges and which can be as deep as 2,000 metres. The greatest heat flow from the ocean floor takes place near the summit of the mid-oceanic ridges. There are also far more earthquakes on and around the ridges than are experienced elsewhere, showing these to be geologically active areas.

An increase in understanding of the Earth's magnetic field led to the realization that periodically this reverses. Such fluctuations can be detected with a device called a magnetometer. It was discovered that, either side of the mid-oceanic ridges, it was possible to detect these past reversals in the Earth's magnetic field. The conclusion was that new material was constantly being thrown up on the ridges and was being pushed outwards on either side. The reversals of the magnetic field demonstrated that this process was ancient but that it was still taking place.

Also of interest were 'deep-Sea trenches'. The trenches are generally long and narrow and they are often associated with, and parallel to, continental mountain ranges. In addition they run parallel to the ocean margins. There is great seismic activity associated with the deep-sea trenches, indicating that they too are associated with the process of seafloor spreading and that they are directly related to the oceanic-ridges.

What is now thought to be happening is as follows: underneath the Earth's outer crust is the asthenosphere. This is a malleable layer of heated rock. It is kept hot because of radioactive decay in elements such as uranium. The source for the radioactivity, which also includes thorium and potassium, lies deep within the planet. The asthenosphere, constantly heated, rises to the surface, pushing new material out at the mid-oceanic ridges. Magma escapes along the cracks formed at the ridges, forcing the new seafloor in different directions. The new material spreads outwards until it makes contact with a continental plate and will then be 'subducted' beneath the continent. The lithosphere at this point sinks back into the asthenosphere, where it once again becomes heated.

Few experts disagree with this basic explanation, partly because it can be seen at work. India, for example, started its life on a completely different part of the planet. It is now being forced up into the body of Asia and the Himalayas are the result – a huge mountain range forced up by the pressure of the two landmasses meeting.

The whole process is known as plate tectonics and scientists were keen to see whether or not a similar process was taking place on the other terrestrial–type planets in our solar system – Mercury, Venus and Mars. Probes sent to these planets have now shown conclusively that plate tectonics do not take place on any of our companion worlds, making it a strictly Earth-bound phenomenon, at least as far as our own solar system is concerned.

This is something of a puzzle. What is taking place in the Earth system that is so different from the other Earth-like planets? What caused plate tectonics to commence in the first place and what is the engine that keeps driving the process? There is a growing body of evidence to show that in both cases the answer is almost certainly the Moon. What is more, it is now being suggested that without plate tectonics the

Earth may not have proved to be a suitable haven for life at all.

Dr Nick Hoffman, a geophysicist at the Department of Earth Sciences, Melbourne University, Australia, has recently suggested that the Moon made plate tectonics happen simply by coming into existence.

As we have discussed, the origin of the Moon is still shrouded in mystery, no matter how much proponents of any specific theory of its origin may pretend. However, there are certain facts that are known for sure. As we have seen, the Moon is definitely made of the same stuff as the Earth, but not all of the Earth. Rather the composition of the Moon closely resembles the material in the Earth's crust, without many of the heavier components, such as iron, that make up the Earth's core.

But how could such a large amount of the Earth leap from the planet's surface into a position tens of thousands of miles in space?

Scientists were puzzled. And then a potential explanation was put forward in the form of the original Big Whack theory – the suggestion that maybe some object, about the size of Mars, collided with the young Earth and that the Moon was formed from surface material that was blasted off the face of the infant Earth. There did not seem to be any other possibility, so it is now regularly taught as though it is a fact. The major problem of the Earth's current speed of rotation was tentatively explained away by proposing a second impact from the opposite direction occurring quite soon after the first.

To us this sounds like a rather desperate scenario to believe in. And as we have seen, other problems remain for this would-be explanation; not least the question of where the material from the incoming objects went to. If the Double Whack theory as correct, the Moon should be made up of three different sets of material, but it is not. It is made of Earth rock alone.

Nick Hoffman, as an acclaimed expert on the terrestrial planets within our solar system, has suggested that the removal of the material that went to make the Moon may have triggered plate tectonics by creating the space for the planet's skin to shift. He points out that on Venus, for example, the same sort of forces are at work but the crust of the planet is so thick, the stresses within the crust simply cancel each other out, with the exception of a few wrinkles here and there. Hoffman has noted that if the seventy per cent of Earth crust that was destined to become the Moon was returned to the Earth, it would 'fill the ocean basins with wall-to-wall continent'.

What would the Earth be like without plate tectonics?

Hoffman suggests it would be a water world, covered with oceans and with only the tips of extremely high mountain ranges poking out above the surface of the water. Of course there is nothing to suggest that life could not have existed on such a planet and Hoffman agrees that life is most likely to develop in a watery environment. It's a fact, though, that what we term as being 'intelligent life', such as our own species, has developed on land. The use of fire would not be possible in a watery habitat and the use of tools, one of the factors that is generally accepted as the starting point of our advance, is also a dry land phenomenon.

In any case, as we will see, the Moon is so important in other ways that even a watery world may have proved to be impossible without its existence.

CHAPTER SEVEN

The Incubator of Intelligence

Nick Hoffman's suggestion that the creation of the Moon removed so much material from the surface of the Earth that plate tectonics could become a reality is fascinating. It is estimated that seventy per cent of the primordial crust of the Earth would be necessary in order to create the Moon. Its removal caused the remainder of the crust to spread, allowing continental drift to take place.

Whether or not this is the whole story, plate tectonics are a reality as far as the Earth is concerned and what is more, it is a phenomenon that only occurs on the Earth. In other words, no other terrestrial-type body in the solar system had continents travelling about its surface.

One of the three Earth-like planets in the solar system, apart from the Earth itself, is Mars, which is half the size and a tenth the mass of our planet. It has an atmosphere that is ninety-five per cent carbon dioxide and nearly five per cent nitrogen with a pressure at the surface that is only 1/200th that of Earth.

Unfortunately for any potential Martian life form, liquid water cannot exist at the ambient pressure and at the temperature of the Martian surface. On this planet, water goes directly between solid and vapour phases without becoming liquid at all.

The puzzle as to why plate tectonics have either never started or else never been maintained on Mars has not been totally explained, but there are theories.

Mars has no appreciable mountain ranges, though it does have giant volcanoes. Some geologists suggest that the absence of true mountain ranges gives one clue as to why Mars did not develop plate tectonics. Like Earth, Mars has a lithosphere. This is a region in the crust of the planet that is cooler than its interior – a little like the skin that forms on a cup of hot milk. The centre of the Earth is extremely hot, probably more so than that of Mars, but the presence of volcanoes on Mars must indicate a hot core. One difference might be that Mars has nowhere near as much water in its composition as Earth. It is thought that it is water trapped within the Earth which acts as a lubricant allowing different parts of its rocky surface to slide against each other. The limited amount of water on Mars seems to prevent the lithosphere from allowing fresh material from deep within the planet to rise to the surface in the way it is constantly doing on Earth. As a result the lithosphere has not been disturbed for aeons and has cooled gradually, getting thicker and thicker. When pressure has become so great within the body of Mars that it is powerful enough to escape, it has done so via volcanism and not along features like the mid-oceanic ridges on Earth.

The other Earth-like body, Venus, which orbits closer to the Sun than our own planet, has a surface very different to that of Mars or the Earth. In some ways Venus is more like Earth than Mars. Venus is a similar size and mass and is also compositionally quite like Earth – or at least it was once. Experts such as David Grinspoon, a research scientist at Southwest Research Institute in Boulder, Colorado, have studied Venus closely, aided by a whole series of orbital and lander space missions.

Grinspoon is not alone in believing that in its early stages of development Venus was even more like the Earth. There is no discernable water on Venus now but there are traces in the atmosphere, which most likely indicates that in its very early stages it had proportionally as much water as Earth. This is not too surprising because the planets formed at the same time and fairly close together.

Venus is not unlike Mars in many ways but its surface pressure is ninety-two times that of Earth. It is thought that Venus lost its water because of a greenhouse effect and it is now covered in dense swirling clouds of sulphuric acid. These clouds are so thick that only a small percentage of the sunlight that falls on Venus actually gets through to the planet's surface, so even if it weren't such a hell in other ways, it would be a very gloomy world. It might be thought that less sunlight would lead to a lower temperature but this isn't the case. Rather, heat already at or near the surface is maintained and increased because it cannot escape through the dense carbon dioxide. This has caused a dramatic heating of the surface of Venus to a present temperature of 730°C.

Like Mars and Earth, Venus has volcanoes; in fact it has more than any other planet in the solar system. But again, like Mars, the volcanoes of Venus exist as individual entities and not as part of long mountain ranges as is the case on Earth. The volcanoes of Venus are randomly spread about its surface and many of them look very recent, even though this may not be the case. Electrical storms rage constantly through the clouds of sulphuric acid but, even so, wind erosion on Venus is limited compared to the water-rich Earth. It turns out that erosion is extremely important in terms of supplying the right chemical and nutrient balances that have made the Earth a haven for life.

The surface of Venus looks broadly similar wherever one looks and is thought to be comparatively recent in origin – something in the order of 600 to 700 million years. Venus has a generally smooth surface with some rifts and folds but everything appears to be the same age. It is generally accepted that between 600 and 700 million years ago some cataclysm took place on Venus that remodelled its whole surface. Whether this was as a result of the internal stresses within the planet is not known, but for some reason the planet's surface appears to have literally melted or more likely was uniformly covered with volcanic basalt.

Nobody knows for certain whether a similar thing will happen again on Venus, in other words whether we are seeing only one phase of a stop-start process that is taking place, but it is considered to be a distinct possibility. Probably because of its greenhouse atmosphere Venus is deficient in water and so once again may have built up a thick lithosphere. It certainly does not display any of the characteristics of plate tectonics.

It is interesting to note that Venus has no moons, whilst Mars has two, though both of these are extremely small and can have little or no effect on their host planet. As we have seen, it is now being suggested that the very creation of such a large moon as that enjoyed by Earth was directly responsible for the start of plate tectonics, which in turn allowed life to form on the planet in the first place.

In the early stages of its existence, the Moon was very much closer to the Earth than it is today. And it is the existence of the Earth's oceans that is primarily responsible for the gradual lengthening of the distance between the Earth and the Moon. This is a process that has been taking place for the last four billion years and which is still taking place.

One way of looking at the situation was presented by Neil F Comins, Professor of Astronomy at the University of Maine. Back in 1990 he had been struck by the comments of a colleague, to the effect that science educators are always looking at the world from the same old perspective. Comins suggested that it might be sensible to step aside and look at the world differently.

As a result of this conversation Comins decided to turn his attention to something we all take for granted, namely the Earth and its relationship to the Moon – but from an entirely different perspective. He set out to consider what the Earth would have been like today if it had not enjoyed the benefits of so large a Moon. He called his hypothetical world 'Solon' and over a period of time he wrote a series of articles about Solon that appeared in *Astronomy* magazine. He eventually published his overall observations in a book, which was entitled *Voyages to Earth that Might Have Been.*[21]

Comins examined every aspect of the Earth and its relationship with the Moon to build a picture of a similar planet, at the same distance from the Sun and which was the same age as Earth. The only thing that was different is that the Moon did not exist, but the alterations this absence would make to the Earth were dramatic.

Nick Hoffman suggests that the very nature of the Earth's surface would have been entirely different if the material that makes up the Moon had not been removed from the Earth's crust. However, Comins' starting point is to assume that the surface details of the Earth would be roughly the same as they are now.

One of the greatest differences in terms of the early, developing Earth would have been tides. Comins makes the point that a Moon ten times as close would have led to daily lunar tides that would have been a thousand times greater than they are today. Bearing in mind that it is generally accepted that the infant Earth was spinning about its centre every six hours, this means that tsunami-strength tides would have been hurtling across the Earth every three hours! Not only were these tides more frequent, but, being so very much larger, they would have crashed many hundreds of kilometres inland – and with tremendous destructive force.

The mechanism that has slowed the Earth's spin is directly related to tides and the Moon is not the only body responsible for them because part of the ocean tides on the Earth are responsive to the Sun. But the Moon is much closer and has done far more to slow the Earth than has the more distant Sun. Comins estimates that without the Moon, the Earth day would be only eight hours in length and solar generated tides alone would be less than a third of what they are today.

The immediate implication has great ramifications on the possibility of evolving life. At present many scientists accept that DNA, the fundamental building block of all life, occurred spontaneously in Earth's early oceans. We will have much more to say about DNA later, but for the moment we will accept the general view that it first appeared in the early oceans of the Earth, a legacy of what is known as the 'primeval soup – a specific blend of water and chemicals upon which life depends.

The massive tides created by the infant Moon would have caused erosion on a scale quite beyond our experience today. Millions upon millions of tonnes of land would have been pulverized and swept out to sea, then widely distributed and eventually settled on the seabed. This process liberated vast amounts of minerals into the oceans – minerals that emerging life simply could not do without. Presumably a Moonless world would still have had weather patterns, including rain, so erosion would have taken place but on a tiny scale compared with what happened when the Moon was so much closer to the Earth. This means that life would have taken much longer to gain a foothold, if it had managed to do so at all.

We have no problem with the concept that life first developed and flourished in the ocean, but there had to be a time at which it migrated from its salty environs and learned to survive on dry land. It is possible that insect life took the leap first but the fish ancestors of amphibians and reptiles followed and between them they eventually gave way to all land-living animals in the world today.

Life is always evolving to harmonize with the prevailing environment and to capitalize on new niches that are not already being exploited. Around 400 million years ago one such area of potential exploitation was rock-pools. Fish are accidentally left behind in rock pools with every retreating tide, both then and now. In most cases it doesn't matter because the next high tide will free the fish again, back into the sea. However, if a fish is isolated in a rock pool during a particularly high tide, it may have to survive for weeks before it will be liberated. Fish that found themselves in this situation would die unless they somehow managed to get back to the ocean by moving over dry land and also managing to breathe out of the water.

It seems that some fish did find ways to drag themselves across the sand, at the same time changing enough physically to take gulps of air whilst out of the water. These fish found that dry land offered some rich pickings and any animal that learned to live, even temporarily, on dry land, would be well rewarded. Gradually, and over a long period of time, fins that pushed the fish over sand became stouter until they became legs and the fish in question ceased to be fish at all.

Since the Sun also creates tides it isn't out of the question that fish would ultimately have left the oceans, even if lunar tides had not been present. However, the waves in question would have been significantly smaller and their value in terms of depositing detritus much more limited. What is quite clear is that life would also have been very much slower in developing to a stage advanced enough to leave the oceans had it not been for the lunar tides, if it could ever have happened at all. When we take on board the prospect of an Earth with a variable obliquity, no plate tectonics and such a dizzying spin about its axis, the prognosis for life of any sort on Comins' Solon is not good.

Fortunately for us the Moon was present and stamped its authority on the developing Earth in a number of different but equally crucial ways. It helped to create many differing habitats, which in turn engendered biodiversity. Most experts believe that it was biodiversity that led to intelligent life becoming possible. Evolution tries and retries many different models. Animals that were ideally suited to their environment flourished on the Earth, only to fall by the wayside when conditions changed and they could not adapt.

Giant reptiles, that we generically call 'dinosaurs', ruled the Earth for millions of years until these impressive and diverse creatures vanished from the face of the planet. Whether as a result of some cataclysm, such as a huge meteorite strike, or thanks to some other misfortune, species that had flourished for eons were wiped out astonishingly quickly, but life itself remained untouched. Such was the multiplicity of species already inhabiting the Earth that some were bound to overcome the problems that put paid to thousands of others at a stroke.

One of the animals that did survive whatever circumstances put paid to the dinosaurs was a tiny shrew-like creature that occupied the vacant niche left by the demise of the reptiles. However, it was different to the reptiles because it gave birth to live young and suckled its infants with milk created from its own body. These first mammals then evolved to diversify and spread across the planet where they have been adaptable enough to survive and flourish.

Tree-dwelling species became monkeys and some of these creatures came down from the trees and began to move across the open savannah, most likely created by yet more climatic changes. Down on the ground these anthropoids were vulnerable. If they were going to survive they were going to need something that had not been specifically necessary to earlier creatures.

They needed bigger brains.

Evolution responded and a whole family of hominids was the result, of which *Homo sapiens* is now the only surviving example. But despite our general sense of specialness, recent events point to our solus position as being surprisingly recent.

One of the greatest breakthroughs for humans was the control of fire; but the earliest known evidence of regular fire using is unequivocally attributed to our larger-brained cousins, the Neanderthals, some 200,000 years ago. We coexisted with these people until they finally disappeared in southern Europe around 25,000 years ago. Science had believed that an earlier hominid, *Homo erectus,* had become extinct hundreds of thousands of years ago, until the mid-1990s when remains found on the island of Java in Indonesia were found to indicate that they too were around until 25,000 years ago.

Both these alternative humans disappeared at a time when midsummer's day fell around June 21st in the northern hemisphere – just as it does today. The dates on which astronomical events such as the summer and winter solstices and the spring and autumn equinoxes fall, move backwards through the calendar by one day (around one Megalithic degree) every seventy-one years. This is due to the long, slow wobble of the Earth on its axis, known as 'the precession of the equinoxes' which takes 25,920 years for each cycle.

This movement through the calendar has no effect on people at all, but it is interesting to note that a recent discovery suggests we were not alone as a species as recently as 13,000 years ago, when the summer solstice in the northern hemisphere fell in late December; the exact opposite of where it is right now.

The discovery of what is claimed to be a previously unknown branch of hominid occurred on the island of Flores, near Java, and was announced to the world in 2004. Remains have been found of a dwarf hominid, named *Homo floresiensis,* which was only as tall as a modern three-year-old with a facial morphology very different to *Homo sapiens.* Strangely, these miniature people had mini-brains yet they produced relatively sophisticated tools.

Not only have we recently shared the planet with other hominids, it now seems that the ancestors of today's Europeans may have interbred with other types of human in the not too distant past.

As part of a large-scale gene-mapping programme, researchers at deCODE Genetics in Reykjavik, Iceland, were looking at the families of nearly 30,000 Icelanders. They found that women who had an inversion on chromosome 17 had, on average, 3.5 per cent more children than women who did not. Kari Stefansson, deCODE's chief executive, considered this to be a very significant impact in terms of an evolutionary timescale. It is possible to roughly date the origin of this phenomenon by counting the number of genetic differences that have accumulated in it compared to a normal DNA sequence. It turns out that this element has so many differences that it must have occurred about three million years ago. Which is long before modern humans evolved.

Stefansson has suggested that this element of the DNA might have been native to some other species of early human and came to our own species around 50,000 years ago. He added: 'There aren't all that many ways you can explain it except by the reintroduction into the modern human population ... That raises the possibility it was reintroduced by cross-breeding with earlier species.'[22]

But as these other humans disappeared, *Homo sapiens* developed a growing intelligence that allowed us to begin to manipulate the environment in which we live. The great breakthrough was the development of agriculture – a move that allowed civilization to emerge.

With civilization came the ability to count and ultimately a way of expressing language in a written form. Knowledge that had once been laboriously passed from one generation to the next could now be stored and retrieved from places outside the human brain. Intelligence also created technology and a great desire to understand the workings of the world and the cosmos of which it was part. But this curiosity began long before we sent representatives of our species to walk on the Moon. It had been present for more than 30,000 years, when the first lunar calendars were created. It is almost certain that after the Sun, the Moon was the most important heavenly body to captivate our species.

How little those cave dwellers, who scratched their knowledge of the lunar cycle onto animal bones and antlers, were aware that without the presence of the lunar disc that so captivated them, the Earth would probably be a lifeless rock, silently spinning around the Sun, like the inferno of Venus and the frozen wastes of Mars.

CHAPTER EIGHT

External Intelligence

'Rather than transmitting radio messages, extraterrestrial civilizations would find it far more efficient to send us a "message in a bottle", some kind of physical message inscribed on matter. And it could be waiting for us in our own backyard.'
Professor Christopher Rose of Rutgers University, New Jersey & Gregory Wright, a physicist with Antiope Associates, New Jersey

The idea that intelligent life forms might exist elsewhere in the cosmos is a comparatively recent interest for humanity. For thousands of years and across countless cultures, it was more or less accepted that anything dwelling outside our own immediate environment inevitably fell into the classification of a god or a servant of the gods, such as the saints, angels or seraphim that inhabit the heaven of the Judeo-Christian tradition.

Even after the telescope appeared, around the year 1600, the Catholic Church in particular was not keen to have its dogmas regarding the nature of the Earth and its relationship with space tampered with in any way. In Christian doctrine, the Sun and the Moon have both been directly created by God, as have the stars and planets. The first book of the Bible, Genesis, lay down the order in which God created the observable cosmos and anyone who seemed to be throwing a spanner in the works, for example Galileo (1564–1642) who suggested that the Sun, and not the Earth, was the centre of the solar system, was liable to be severely censured. Galileo was forced to recant his heretical views and was condemned to perpetual house arrest but was probably lucky to escape with his life.

Even before Galileo's time, thinking people were not fooled by the Church's account of the solar system. The Portuguese navigator Ferdinand Magellan (1480–1521) understood what he was seeing at the time of a lunar eclipse: 'The church says the earth is flat, but I know it is round for I have seen its shadow on the moon and I have more faith in a shadow than the church.'

Only the effects of the Renaissance and Church reformations across Europe broke the hold of old church dogma. By the late seventeenth century, with telescopes proliferating and almost anyone able to take a close-up view of the Sun, Moon, planets and stars, the cat was truly out of the bag and the genuine nature of the solar system in particular was beginning to become apparent.

Since Charles Darwin wrote *The Origin of Species* in the mid-nineteenth century it has become clear that life on Earth has evolved over billions of years from the first single-cell entities through to all of the creatures in the world today. Darwin's ideas were argued over fiercely at the time, but the massing evidence from palaeontology, genetics, zoology, molecular biology and many other fields gradually established evolution's truth beyond reasonable doubt.

It is ironic, therefore, that the most scientifically advanced nation the world has ever known, the United States, has large numbers of 'Creationists' – people who still cling to the teachings of the mediaeval Church. They are currently trying to persuade politicians, judges and the general public that evolution is an unproven myth cobbled together by atheists. They lobby for their ideas, such as 'intelligent design', to be taught as alternatives to evolution in science classrooms. Their proponents admit that their aim is to keep the scriptures of the Christian religion taught in school as the word of God, rather than a collection of ancient Jewish texts.

Their arguments against Darwin's concept of 'natural selection' are not well reasoned or based on any normal principle of modern science. These people appear to be intellectually stuck, hundreds of years in the past, at a time before masses of new data became available. However, it is interesting to note that academics once thought like this too. Dr John Lightfoot, the Vice-chancellor of the University of Cambridge was not frightened of being precise about the origin of the entire Universe when he said in 1642:

'Heaven and earth, centre and circumference, were created together, in the same instant, and clouds of water ... This work took place and man was created ... on the 17th of September 3928BC at nine o'clock in the morning.'

Poor Dr Lightfoot seems to have been ignorant of even the most basic facts of science. He clearly did not realize that there is no such thing as nine o'clock in the morning because every hour of the day exists simultaneously on our revolving planet; it just depends where you are standing. Happily, the very year that Lightfoot made this statement, a baby boy was born in the village of Woolsthorpe in Leicestershire. The infant's name was Isaac Newton and he went on to become Cambridge University's most famous professor and a man that would create a leap forward in humankind's understanding of the Universe.

Newton however, did not dismiss the role of God as he wrote on Judaeo-Christian prophecy, the decipherment of which he saw as being essential to the understanding of God.

His book on the subject espoused his view that Christianity had gone astray in 325AD, when the crumbling Roman Empire declared that Jesus Christ was not a man but an aspect of the very deity that had built the Universe.

Today we have the benefit of masses of data from all kinds of disciplines that point to the Earth being nearly five billion years old, but many creationists frequently quote the chronology produced by James Ussher who was Archbishop of Armagh and Primate of All Ireland in the early seventeenth century. His analysis, based on his interpretation of the King James Bible, allowed him to confidently declare that the creation of the world occurred in 4004BC.

Such a dating raises all kinds of problems, from fitting in the obvious existence of dinosaurs, for example, to the fact that the city of Jericho, near to the River Jordan, has been continuously occupied for 10,000 years. (Interestingly, the origin of the name 'Jericho' is Canaanite and means 'the Moon').

There are creationist websites that put forward 'evidence' that their writers believe demonstrates that people and dinosaurs lived at the same time – presumably around the time that the Megalithic Yard was being introduced! But these are not fringe ideas as there are large numbers of people who believe that geological time is a myth. According to a survey run by the Gallup Organization in 1999, the majority of Americans educated up to high school level or less, believe that God created humans in their present form within the past 10,000 years or so. And a worrying forty-four per cent of college graduates believe the same.

An international research team led by scientists at the University of British Columbia sees the creation as being a little earlier than Dr Lightfoot and Archbishop Ussher. Professor Harvey Richer, the study's principal investigator, confirmed previous research that sets the age of the Universe at thirteen to fourteen billion years. The team measured the brightness and temperatures of white dwarf stars (the burned-out remnants of the earliest stars which formed in our galaxy) because they are 'cosmic clocks' that get fainter as they cool in a very predictable way.

More recent calculations, by Lawrence Krauss of Case Western Reserve University and Brian Chaboyer at Dartmouth College, published in the journal *Science,* put the Universe at anything up to twenty billion years old.

Creationists often try to invalidate all of evolution by pointing to science's current inability to explain the origin of life. John Rennie, the editor in chief of *Scientific American* has countered this by saying:

> '...even if life on Earth turned out to have a nonevolutionary origin (for instance, if alien's introduced the first cells billions of years ago), evolution since then would be robustly confirmed by countless

micro-evolutionary and macro-evolutionary studies.'[23]

It is true that, whilst science can explain how life has evolved on Earth, the way it all began is a complete mystery. And, as far as we know, the Earth is the only location where life exists.

In the nineteenth century some people speculated that there might be life, or even people, living on the Moon. It is now certain that no natural life could exist on the Moon, which is a barren world constantly irradiated by the Sun and lacking in both available surface water and a sufficiently dense atmosphere to support life. There was a more recent time when Venus, the second planet out from the Sun, seemed a potential candidate for some type of life because its dense clouds hid the surface from view so that, for all we knew, it might be as green and verdant as that of the Earth. But as we now know, it is furnace hot and continually subjected to sulphuric acid rain. As a result, the chances for life seem almost nonexistent.

Mars is certainly cooler and there may be water existing near its polar regions. At the time of writing this book, some people are still clinging to the possibility that there could be some sort of primitive life on Mars either now, or at some time in its remote past. If it does exist at all, life on Mars is likely to be extremely simple. Other planets in the solar system, being gaseous giants in the main, are even less likely to support any sort of life as we know it.

By far the majority of experts now accept that if advanced life of any sort does exist in places other than the Earth, we will almost certainly have to look deep into interstellar space to find it. Our solar system is only one of many that undoubtedly exist, even in our own corner of space. Astronomers have identified suns that have planets orbiting them and it is estimated there are a thousand million stars in our own galaxy, any one of which could possess a planetary system where life might have evolved and flourished. Beyond our galaxy there are countless others, so it may be wrong to think that only our tiny little blue planet, amidst such a proliferation of planet-bearing suns, has produced a thinking species such as our own.

But as far as we know right now, we are alone.

Once the sheer size of space was ascertained it also became apparent that even if there are hundreds or thousands of intelligent species out there, the chances of us actually encountering them in any way is quite small. Distance is a problem but it isn't the only one. One of the greatest stumbling blocks could be time itself. In order for us to communicate with another advanced species, it would have to have reached at least our level of sophistication either at the same time as us or shortly before. Although humanity has created at least a couple of probes that are presently leaving the environs of our own solar system, it will be decades, or maybe centuries, before we embark on interstellar space travel to any significant extent. Even if we do, the answers we are looking for, in terms of finding other intelligent beings, are likely to be protracted.

The thought of any spacecraft travelling faster than the speed of light remains in the realms of science fiction. If, as Einstein proposed, light speed is as fast as anything can ever travel, it would take many years merely to reach the nearest star. To go beyond our own galaxy, the Milky Way, would seem impossible because the next nearest place we could visit is the Sagittarius Dwarf galaxy which has 'only' a few million stars and is a staggering 80,000 light years away. The next nearest galaxy is the Large Magellanic Cloud and that is 170,000 light years distant.

Setting out to actually meet our intergalactic or extragalactic cousins seems to be a hopeless idea, even if we knew where they were located. So does this mean we can never say hello to any of them? Not necessarily. If we cannot greet them face-to-face, it might be possible to listen to them.

Much of the energy so created streams out into space as electromagnetic radiation. There are many wavelengths of this radiation, some of which are familiar to us in our daily lives. The full panoply of this radiation is known as the 'electromagnetic spectrum'. The shortest of the wavelengths are those we call 'gamma waves'. At the other end of the electromagnetic spectrum are extremely long radio waves, which we harness every day. Visible light is also a component of the electromagnetic spectrum, as are the microwaves used daily in many cookers.

In fact we are getting radio messages from all parts of the cosmos all the time. These are emitted by suns and other much stranger bodies within our own galaxy and beyond it, as a result of the physical processes taking place within them. Electromagnetic radiation travels across the near vacuum of space at the speed of light. Once it was realized that we could listen in on the processes taking place in our stellar backyard and beyond, radio astronomy was born.

In 1931 an American engineer by the name of Karl Jansky, who was working for the Bell Telephone Laboratories, was conducting experiments into interference that was taking place across certain radio wavelengths. He built a succession of aerials and managed to isolate three distinct sources of radio interference or static. Firstly he could detect local thunderstorms; and secondly, storms taking place at a greater distance. However, there was a third source of interference that was steady and always present which he couldn't, at first, identify. By moving his aerials, Jansky was eventually able to isolate the source of this third form of radio interference. To his own and many other people's great surprise it was coming from within the Milky Way and in fact it originated at the very centre of our own galaxy.

Like many controversial discoveries Jansky's were ignored for some years. But not everyone was sceptical. Reading about Jansky's observations, in 1937 another radio engineer, Grote Reber, built his own aerial, though this one would have been more familiar to a modern radio astronomer because it was a dish. Reber also picked up the strange 'messages' from space.

Interest in the signals from space gradually increased. In 1942 a British Army officer, J S Hay, made the first observations of radio emissions from our own Sun, whilst working on ways to jam German radio signals. Once the Second World War was over, radio astronomy really took off and within a few years discrete signals from all parts of space were being received. Ultimately a background radio source was recognized that could not be isolated to a particular point in space and it was finally realized, in the 1960s, that this was the signal left by the Big Bang – the very birth of the Universe itself.

Of course, all the signals that were being received were perfectly natural in origin. But towards the end of the 1950s it began to occur to a number of those involved in radio astronomy that if any species out there in space was already more advanced than we were, it might well make use of radio waves in order to let us know it existed. Most radio signals received from space can be readily identified and even those that proved to be a puzzle at first have been shown to have a natural origin. But if an advanced species actively wanted to send a message, it would not be difficult for it to use a type of radio signal that could not be confused with that created by any natural phenomena – for example, one containing an obvious mathematical formula.

In 1961, when the 'race for space' had fired the imagination of a generation, a new organization came into existence. It was called SETI – 'the Search for ExtraTerrestrial Intelligence'. SETI was primarily the brainchild of an enthusiastic young electrical engineer turned radio astronomer by the name of Frank Drake, a 31-year-old engineer who had become interested in radio astronomy whilst at Harvard Graduate School.

Drake was fascinated by the prospect of radio astronomy being used to identify other intelligent species in the cosmos and thought that we should be actively listening in for any message that might be transmitted from deep space. Together with another interested scientist, J Peter Pearman, an officer on the

Space Board of the National Academy of Sciences, Drake arranged the first SETI conference.

Anxious to show the world just how likely extraterrestrial life surely was, Drake came up with what is now known as the 'Drake Equation'. This reached the conclusion that there must be many thousands of intergalactic civilizations capable of creating and sending radio messages across space.

The idea of SETI was immediately popular with the public and for a while NASA had some involvement. During the 1960s and '70s, NASA's contribution was fairly low-key, but in 1992 NASA initiated a much more formal SETI programme. Unfortunately, less than a year later, the United States Congress cancelled the funding and NASA, reluctantly, pulled out of the SETI research programme. This certainly wasn't the end of the story because a proportion of the intended NASA research was taken over by the non-profit-making SETI Institute and by an associated body, the SETI League.

SETI has now enlisted the help and support of people from around the globe. Many computer users are regularly sent packages of information received by SETI, in order that it can be analyzed during computer down time. Millions of individuals are involved in what is known as the SETI@home project at the present time.

Exactly where in the electromagnetic spectrum we should be listening for deliberately created messages from the stars was decided in 1959. Phillip Morrison and Giuseppe Cocconi, two young physicists at Cornell University in the United States had co-operated to submit an article to the prestigious science journal, *Nature,* which appeared in September 1959. It was entitled 'Searching for Interstellar Communications'. When trying to ascertain which part of the electromagnetic spectrum to monitor for alien signals, Morrison and Cocconi ultimately opted for a frequency of 1420MHz. Not only does this frequency fall in a very 'quiet' part of the available spectrum, it also represents the emission frequency of the most common element in the Universe, which is hydrogen. Morrison and Cocconi believed that any intelligent species would realize these two facts and so would therefore be most likely to transmit a greeting at or around this frequency.

Some promising messages have been received across the last three decades but, in the end, all of them turned out to be natural phenomena. Space can supply some surprisingly 'ordered' signals. Rapidly spinning objects in space known as 'pulsars' are a good case in question, so SETI experts are extremely careful and also deeply sceptical when any apparent 'letter from the stars' is announced.

One of the greatest problems for SETI, or indeed anyone trying to pick up a message from space, is knowing exactly what to expect. It is certain that any species sending such a message will be in advance of us technologically because if the message received comes from deep space it must have taken thousands or even millions of years to reach us. The culture that sent it might, by the time it is received, have disappeared, advanced even further or simply become bored with the whole notion. All we can do is to take an educated guess and suppose that for any species there will be commonality in terms of the irrefutable laws of physics.

We may receive a logically repeating mathematical sequence such as pi or a list of prime numbers, it is simply impossible to know. There are sceptics around who suggest that the whole process of looking for such a message is destined to fail, if only because other intelligent species out in space may be so different to us that there would be no points of contact recognizable on both sides. In other words, they may be trying to contact us right now and we simply cannot understand the message.

By the summer of 2004 we were already beginning to reach our own conclusions about how an intelligent species from elsewhere might have already contacted us – humanity simply had not recognized the fact yet. Serendipity being what it is, an article appeared in the August 2004 edition of *New Scientist.* It was written by Paul Davies, a scientist at The Australian Centre for Astrobiology at Macquarie University, Sydney. We found it pleasing that a respected scientist was publicly discussing the idea that an alien culture may have put a message intended for us in place many millions of years ago: a message, that Professor Davies also likens to the plot of the film *2001: A Space Odyssey.*

Whilst congratulating SETI for its efforts to track down incoming messages from space, Paul Davies makes the suggestion that to try and contact humanity by way of radio signals might prove to be fairly unreliable for any alien species far away. He points out that the problem of 'timing' might make radio contact difficult, if not impossible. No matter how many such intelligent societies there might be, the chance of them transmitting during the short time slot during which we have been listening is very remote. Is it not possible, Davies asks, whether such a culture, probably immeasurably older than our own, may have conceived of a much more reliable way to let us know of its existence?

Might it not have opted for a method of communication that was not dependent upon transmitting signals for many millions of years in the hope that we, or someone like us, had just evolved the ability to decipher messages in the form of radio waves? Would it not be more likely that our intergalactic cousins would have chosen something much more timeless?

This suggestion, when we read it near the start of Davies' article, made us sit upright and pay attention because we were already asking ourselves the same question. Davies goes on to suggest that, rather than radio messages, a far more reliable way for any alien species to contact us would be to leave artefacts in the vicinity of planets likely to spawn intelligent life that, given sufficient advancement on the part of such a developing species, it could not fail to recognize.

Then we came across yet more heavyweight scientists with similar, highly logical, thought.

Professor Christopher Rose of Rutgers University in New Jersey and Gregory Wright, a physicist with Antiope Associates also in New Jersey, have stated that the transmission of a radio signal by an extraterrestrial civilization, that would probably have to be detected 10,000 light years away, does not make sense. They suggest that it would be far more efficient to send us some kind of physical message inscribed on physical matter – a kind of 'message in a bottle'. And, they believe, such a message could already be waiting for us in our own backyard.[24]

Rose observed that: 'If energy is what you care about, it's tremendously more efficient to toss a rock.' Once radio signals pass us by they are gone for ever, so aliens would have to beam signals continuously as we have only had radio for a miniscule fraction of our existence as an advanced species.

We had to ask ourselves, what if that physical object was the Moon and the information is there for us to see – once we understand the vocabulary?

If the Moon does hold a message, it would be exactly what Paul Davies called a 'set and forget' technique that would survive for millions or even billions of years. Any conventional sort of physical structure, no matter how impressive, would eventually crumble under geological forces, especially on a very active planet such as our own. It turns out that the possibilities for a 'letter from the stars' that can survive eons are actually very limited indeed. In the end such a 'physical' message needs to be either extremely large or extremely small – and as we were to discover, perhaps both.

We had already uncovered a wealth of published academic material that points to the Moon being the single most important factor in the development and nurturing of complex life forms on the planet Earth. Quite simply, if the Earth is thought of as an incubator for life – the Moon is the carefully programmed machine that monitors and stabilizes the process. A real life-support system.

This may be a wonderful coincidence of epic proportions, or it could be yet another miracle to ignore as an inevitable consequence of the 'Anthropic Principle'.

Whether or not the suggestion put forward by Davies has any merit, it was the second time in a month during which perfectly respectable and serious scientists had published articles dealing honestly with the possibility that we are not alone in the Universe and that other species may be trying, or might have tried in the past, to contact us. Whilst we are delighted with the open-minded attitude that seems to be developing with regard to this subject it is our profound belief that the message, for which SETI, Paul Davies and Rose and Wright are seeking, is right in front of our eyes. It has been there as long as humanity has existed and what it has to tell us is breathtaking in its implications.

A Potential Message?

At this point we asked ourselves what would a message look like that had been planted on Earth or in its immediate environs and which was intended to survive for a huge period of time. Firstly, we reasoned, it would have to be recognizable, so it could be clearly interpreted as a message before its contents could be deciphered. Secondly, it would need to be either extremely large or else very small in order to survive the destructive power of Earth's geology and weather systems.

In Clarke and Kubrick's *2001: A Space Odyssey,* the technique used by the unknown aliens was to have a major anomaly that could only be detected by a technically competent species. By placing objects with huge mass under the surface of the Moon the aliens knew they would be easy to spot in a place where there was little else to distract. But in that story the purpose of the gravitational anomalies was not to communicate with the new species – it was to send a message back to the aliens that the local creatures had reached a specific level of intelligence.

So, an anomaly of this sort might be enough to alert an unknown species, such as our own, that there is a message waiting. The next step would be for those planting the message to ensure that the target species understood that it was addressed to them.

It has long been agreed that numbers are the best way to communicate with intelligent creatures from another world. Hieroglyphs or any kind of marks are unlikely to be understood without any point of reference, just as there was no way to understand ancient Egyptian hieroglyphs until the Rosetta Stone was discovered in 1799. This inscribed artefact from the second century BC gave the same text in both Greek and hieroglyphs; thereby providing the key to understanding a lost language.

If numbers are used in such a message they need to have a discernable pattern, however they are communicated, so that they stand out against the background 'noise' of number values that surround us everywhere. Even then, it is extremely challenging to think of numbers that are certain to be spotted once they are planted in our own environment. The safest method would be to use ratios that stand out, because ratios are not dependent on units of measurement or any chosen base (e.g. base ten in everyday usage or base two [binary] as used in computing).

But it occurred to us that we were examining this process backwards, because we started out by being alerted to major anomalies specifically related to the Moon. Not only does it appear very unlikely that the Moon could have occurred naturally in the first place, it also turns out that it has been the incubation machine that so perfectly nurtured life.

We now needed to go back to our starting position and look at the numbers that had fallen out of the Earth–Moon–Sun relationship in terms of ratios and to those measurements that stood out so well when we applied Megalithic units to them.

The first and most obviously strange thing about the Moon is how it appears to be the same size as the Sun when viewed from Earth. It is 400 times smaller and 400 times closer to the Earth than the Sun. Assuming for a moment that this might be the first part of a calling card from an unknown source rather than just a bizarre coincidence, we have to note three factors:

1 It is designed to be meaningful only to intelligent creatures living on the Earth's surface.

2 It is designed to be noticed at this specific point in time, give or take a million years or so each way, because the Moon only behaves in the way it does at this time.

3 It appears to be addressed to a species with ten fingers, because the ratio relationship of that between the Moon and the Sun is such a round number when expressed in base ten (e.g. in base eight the ratio would be one to 620).

Now we will speculate that the ratio of the Moon and the Sun just might be pointing to a deliberately created message. In order to do so, we must suspend all preconceptions of what seems sensible and consider instead what under normal circumstances might be judged unthinkable.

We will therefore temporarily accept that points 1, 2 and 3 above are valid and that someone or something is trying to direct the attention of Earthlings, sporting ten digits and living during this particular point in time, to look at the Moon as a potential message.

So, here we go!

CHAPTER NINE

The Potential Message

In the beginning there was neither existence nor non-existence; there was no atmosphere, no sky, and no realm beyond the sky. What power was there? Where was that power? Who was that power?
Rig Veda 10:129.1-7 (circa 4000BC)

At this point we had decided that we could not continue to gather further facts without having some plan in place. The way forward seemed to be to develop a hypothesis so that we could see how the component parts of the puzzle may fit together. We agreed to suspend all negative comments for a time, so that we did not miss a point by rejecting something that challenged our preconceptions. Only when we had a complete model for our hypothesis, would we critically appraise it and compare it to other possible explanations.

So, we are now entering the modelling world by temporarily accepting the following three concepts as real:

1 The Moon was engineered by an unknown agency circa 4.6 billion years ago, to act as

an incubator to promote intelligent life on Earth.

2 The unknown agency knew that humanoids would be the result of the evolutionary chain.

3 That unknown agency wanted the resulting humanoids to know what had been done and they left a message indicated by the dynamics of the Moon.

We were well aware that there were some issues with these assumptions, which, if we forced ourselves to reconcile them simultaneously, would inhibit lateral thinking. One such problem was the issue of motivation: Why would any agency want a grand plan that spanned a period that was equal to around fifty per cent of the age of the Universe at that astronomically distant time? This would be long-range planning beyond all comprehension. Even then, how did such an agency know that the resulting life form would have ten digits? We would try and deal with these issues but it might be necessary to tackle them at a later stage.

Another problem that had to be put to one side for the moment, was the issue of how Stone-Age builders came to be using units of measurement that are the key to decoding the message. These issues, amongst others, will have to be dealt with in due course, but now we will review the basis of the message.

At the root of our hypothesis is the idea that the very dimensions and movements of the Moon are designed to alert us to the fact that this is not a natural body. We therefore need to go back to the beginning of the solar system itself.

No one knows for sure how the solar system came into existence but, despite all of the ideas about the Moon being made from the Earth, everyone agrees that the Sun, Earth and the Moon were all formed around 4.6 billion years ago. It is thought that the Earth and the Moon were formed very soon after the Sun became a star.

Theories come and theories go, but it is likely that it all began when a vast cloud of dust and gas in an empty region of our galaxy became compressed by starlight and gravitational forces, thereby causing an accumulation process. Otto Schmidt first put the theory of 'accretion' forward in 1944 and as more and more evidence has become available, competitor theories have withered away.

In some way, that astronomers do not yet understand, the Sun was formed and produced light and heat much as it does today. The cloud of dust and gas that was wheeling around the new star, kept on cooling and shrinking and whirling faster and faster before separating into rings. Each of these rings also kept on cooling and shrinking and is thought to have gradually gathered together into a sphere of fiery gas, in the case of the terrestrial planets, before cooling so much that the main part of it became liquid and, eventually, solid.

Until quite recently there were two principle theories to explain the Moon's existence. One was that it was an object that had formed elsewhere and was somehow taken into Earth's orbit; and the other was a theory called 'co-accretion' or the 'double planet' hypothesis. This second theory supposed that the Earth and the Moon simply grew together as twins, born out of the primordial swarm of small 'planetesimals'. However, when lunar rocks were brought back to Earth it was realized that the Moon has no substantial metallic iron core and that its rocks have oxygen isotopic ratios that are identical to the Earth's.

The theory that the Moon had originated elsewhere died instantly because it was obvious that the rocks had formed in exactly the same region of the solar system as those of the Earth. But the worrying point was that the only alternative theory was as effectively debunked as the 'capture' hypothesis. The 'co-accretion' could not be correct because the type and proportion of materials should be pretty much the same for both bodies if they were twin planets.

Suddenly there was no theory of the Moon's origin in existence. Scientists tell us that nature abhors a vacuum – but scientists abhor a vacuum even more. Something had to be found to explain the inexplicable.

It took some years, but in 1984 an idea that seemed to explain the facts was put forward. The original Big Whack theory was an attempt to explain how the Moon could be made of selected Earth materials. For reasons already covered, it is a theory that simply does not work and we are still in a position where there is no watertight explanation for the Moon being where it is.

So let us return to what is generally held to be true about the early solar system. The process of making the Earth was not especially quick, as Stein B Jacobsen, a professor of geochemistry at Harvard University says: 'Within 100,000 years of the formation of the Sun, the first embryos of the planets Mercury, Venus, Earth, and Mars had formed ... Some grew more rapidly than others, and within ten million years, about sixty-five per cent of Earth had formed.'[25]

Let us now consider a theory of the Earth–Moon system that does work. The thrust is that it is the result of intelligent design. We do not know who or what this hypothetical intelligence was, so we will designate it UCA (Unknown Creative Agency) for the time being.

In the Beginning

The young star was shining out, and the clouds of matter that had recently circled around it in a series of rings had begun the process of accreting into spheres at differing ranges from the star. One of these proto-planets was some 150,000 kilometres distant from the mother star and the UCA realized that it had the potential to produce intelligence.

The UCA may believe that the Universe is destined to die, maybe by becoming a smooth and static soup of incredibly thinly dispersed matter, just a tiny fraction above absolute zero. A nothingness that would essentially be the end of everything – even time.

The goal of the UCA was to seed life wherever possible, to create intelligent beings that could flourish and go out and seed more life themselves. In this way the very fabric of the Universe would be turned into self-aware matter that would slowly halt and reverse the mindless spiral into entropy and eternal chaos. They had a model to use for this location – one that would produce a specific type of intelligent creature, based on carbon and enabled by liquid water.

But it would take several billion years for the tumbling sphere to stabilize and go through a process of evolution of life forms that would result in a species with the intelligence and, more importantly, the imagination to understand their role of striking out into the cosmos to shape and form swirling stardust – and then give it the spark of life.

It was important that when this fledgling planet spawned its thinking and technologically able offspring, the creatures would understand exactly what had happened to bring them into existence, so that they could eventually repeat the process themselves. In this way, a self-aware Universe would continue to replicate itself across the massive span of space and time.

The engineering requirement was demanding.

The proto-planet was completely unstable and it was destined to develop a surface that would be far too rigid to create the necessary conditions for life to begin and to thrive. It required a regulator – a gravitational presence close by, that would tip it over just enough to cause the surface to have a tiny temperature range that would oscillate gently to evenly distribute the energy radiated from its mother star. This also had to be a regulator that would initially use its gravity to plough the surface so that essential minerals could be released for the life-development process to continue.

It was clear that the planet needed to have a loosened surface and the obvious conclusion was to manufacture the regulator from the surface material. This would reduce the tendency of the surface to form one continuous crust and would allow movement within the crust itself. Judging the required mass, size and orbital characteristics of the regulator was a stupendously complex calculation, because it not only needed to have a changing relationship with the planet over time – it also had to contain the message addressed to the resulting intelligent life form.

The equilibrium point was calculated and it was found that seventy-four quintillion tonnes would have to be removed from the planet to manufacture the regulator. To meet all requirements it was going to need a mass that was only 1.234 per cent of the revised planet yet its physical size had to be a relatively large, being 27.322 per cent of its parent. It would therefore have to be made with the barest minimum of heavy elements such as iron and, even then, it would need to either be partially hollow or have the consistency of a sponge.

Mechanisms were then put in place to remove a stream of material from the young planet, that would be spun into a new planet in close Earth orbit. The design was such that the regulator would slowly increase its orbital distance until it reached an average range of around 384,500 kilometres at the expected time of the arrival of intelligent life. This would mean that to the creatures on the planet's surface, the disc of the regulator would appear, with the naked eye, to be the same size as the star at the centre of the system – which would be the first line of the message to make the developing Earth creatures become curious about the regulator. The realization that the factor for each was precisely 400 would also indicate that the message would be delivered in base-ten arithmetic.

Perhaps the UCA used something like black-hole technology to carefully strip only lighter elements from the infant planet. A black hole is a super-dense entity with so much gravity that even light is jsucked into it, like dust into a vacuum cleaner. A black hole with the mass of Mount Everest would have a miniscule radius – roughly the size of an atomic nucleus – and current thinking is that it would be hard for such a black hole to swallow anything, although it would certainly attract material like a giant magnet. Such an idea might explain the mascons (the regions of high gravity) still found on the Moon.

However the engineering job was undertaken, the biggest challenge was to communicate the message that the intelligent creatures endemic to the Earth would spot, because the very familiarity of seeing the regulator in the night sky would cause them to take it for granted. And the UCA knew that intelligence sometimes leads to a dulling of imagination, resulting in confusion between 'describing' and 'understanding'. Contrary to what some intelligent individuals consider, the ability to describe something does not equate with understanding it.

The next levels of the message needed to be more difficult to ignore. The decision was made to create number patterns that stand out as very strange.

The UCA realized that it needed to draw further attention to the artificial nature of the regulator by building on a truly fundamental number that represented the planet. The number that was chosen was the planet's spin rate per orbit of the star, which would have to be instantly recognizable as a value that was unique to the planet – a natural PIN (personal identification number) for an entire world. So, in this case it was a Planetary Identification Number that was required.

At the required time window, the planet would be rotating at a rate of 366 revolutions for each orbit around its mother star, and the use of the number value 366 would therefore be easily spotted as the Planetary Identification Number.

The intelligent creatures would recognize the PIN number from an early date as it requires only very basic astronomy to appreciate that this three-digit number is the most fundamental of all numbers that are unique to the planet.

Surely these Earth-dwelling creatures would be very surprised when they calculated the relative size of their planet to the orbiting regulator and discovered that the one is 366 per cent larger than the other.

The regulator was also engineered with a PIN number that was meaningful to the intended intelligent creatures. That number would be the reciprocal of the planet's PIN number – the mirror image of 366.

The mathematics was simplicity itself. The regulator's PIN number would be arrived at by considering its size as 100 per cent and dividing it by the relative size of the planet, namely 366 per cent. Working to five decimal places the result is:

$$\frac{100}{366} = 0.27322$$

The regulator was then carefully engineered so that at the key point in time it would be orbiting the planet at a rate of once every 27.322 planetary days.

Surely, the creatures would notice that? And as an extra layer, if they looked at the issue the other way around, the size of the regulator compared to the planet has precisely the same number value – being 27.322 per cent of its parent.

Surely the intelligent Earth creatures could not fail to be alerted by such an unbelievably improbable number matching? There is absolutely no reason why the regulator's orbital period in planetary days should numerically echo the relative size relationship it also enjoyed with the Earth.

The consequence of these arrangements would not be lost on the new life forms because they would easily realize that for every 10,000 of their planetary days, the regulator would complete exactly 366 orbits of the planet. Surely they would spot the use of round base-ten numbers and the PIN number 366 being echoed by the regulator?

But then, if they did not recognize these message patterns it would mean that they still lacked the intelligence or imagination to be considered mature.

Fitting the Moon into the Earth–Sun Model

It struck us as extremely likely that the UCA must have had some control over the Earth's rate of spin and its orbital speed, so that they could ensure that it got to the magic 366 rotations at the required time. From everything that is known about the Earth, its orbital speed has been steadily decreasing for a long time but to the astonishment of scientists at the National Institute for Science and Technology in Boulder, Colorado, it suddenly stopped this deceleration in 1999. CNN reported the story on January 2nd 2004 saying:

'Experts agree that the rate at which the Earth travels through space has slowed ever so slightly for millennia. To make the world's official time agree with where the Earth actually is in space, scientists in 1972 started adding an extra 'leap second' on the last day of the year.

For twenty-eight years, scientists repeated the procedure. But in 1999, they discovered the Earth was no longer lagging behind.

At the National Institute for Science and Technology in Boulder, spokesman Fred McGehan said most scientists agree the Earth's orbit around the sun has been gradually slowing for millennia. But he said they don't have a good explanation for why it's suddenly on schedule.'

This caused us to look up the actual speed that the Earth has settled at in its circumnavigation of the Sun, and we were surprised to find that its mean orbital velocity is almost exactly one ten thousandth of the speed of light in a vacuum. At 29,780 metres per second, the variance is less than two-thirds of one percent.

We thought that this was probably a coincidence – but we could not pick and choose which factors are, and are not, significant. And we had to remember that the value 10,000 had already shown up in the number of Earth days for every 366 lunar orbits.

We next turned our attention to the Sun. The diameter of the Sun is estimated at 1,392,000km and as the average diameter of the Earth is 12,742km, so it follows that 109.245 Earths could be placed side by side along the diameter of the Sun. This is not a number that stands out for any reason – at least not immediately. But when we looked at the number of Sun diameters in the Earth's aphelion (its greatest distance to the Sun) we found that there are 109.267, effectively an identical value because the estimate of the Sun's diameter is within this tiny margin.

How strange. There are the same number of Earth diameters in the Sun's diameter as there are Sun diameters between the Earth and the Sun. This is a near perfect echo that does not work for any other planet in the solar system.

These numbers are ratios and are therefore real and independent of units of measurement. But the number also stood out because there are 10,920.8km in the Moon's equatorial circumference. At the time we noticed this, we considered that it really did have to be a coincidence because the number of kilometres in anything just could not be relevant since the metre is a unit that is an invented human convention.

But then we realized that the Moon turns at a rate of precisely one kilometre every second at its equator and that did strike us as very odd. Maybe we had been too hasty in rejecting the role of the metric system.

Our observations about the patterns inherent in the size and movement of the Moon, in terms of ratios, stand out as being beyond mere accident. Although we accept that the apparent patterns that rely on units of measurement, such as kilometres, are far harder to accept without an explanation of how this could have come about. Any 'reasonable' person would immediately reject such factors as meaningless – but then we think there is a great deal in the old adage that 'all progress is dependent on the unreasonable person'.

And some people would not even get to the point of recognizing the patterns in the ratios within the Sun–Earth–Moon system. A scientifically trained person looking at any one of these points would almost certainly respond by saying that 'all numbers are equally valid'. A value such as 100 or 40,000 dropping out of the mix is just as likely as any other number.

We absolutely agree with this view and we would ignore such results if they were only happening once or even twice, but we are confronting a whole list of non random-looking values that add up to create what would otherwise be the most unlikely series of chance events in the history of the cosmos. And, in our view, anyone who dismisses all of these points as coincidence is being either very illogical or downright dishonest.

It is absolutely true that if someone tosses a coin 100 times and it comes out heads every time, the chance of the next toss resulting in another head is exactly 50/50. However, if this ever happens to you in the real world, we would suggest that, before you let them toss the coin again, you check that it is not double headed. Only a fool would not be suspicious.

Scientific discovery has always been a process of identifying patterns that stand out from the chaos of random events. For example, identifying areas where there are more cases of a specific illness is likely to point to a local factor such as radioactive bedrock, a leakage of harmful industrial effluence or a contaminated food plant. When something varies markedly from the norm there is usually a reason.

If we look at the available information logically, and without preconceptions of what is and is not possible, the Moon appears to have been inserted into the Sun–Earth relationship with the accuracy of the proverbial Swiss clockmaker!

A Recent Interaction

It appears that no one has previously spotted this message and we only came across it because of our findings relating to Megalithic units. We suspect that the problem is one of too much knowledge and a loss of the ability of experts, in our super-technological world, to think simply. Perhaps if Galileo or Isaac Newton had access to the information that we have today they would have noticed these issues concerning the Moon, but, alas, they did not have the accurate measurements that we have today and therefore they could not observe the patterns. Today we have the necessary information, but astronomers are understandably more interested in quasars, pulsars and all kinds of deep space objects rather than the fundamentals of the Earth–Moon relationship.

We now needed to consider a scenario that would explain how the Megalithic Yard came to be involved in this ultra-long-distance message.

Maybe the UCA was aware of the potential problem of the message headline being missed due to over-sophistication and took steps to inject extra information near to the key moment when the message needed to be interpreted. Perhaps, we mused, the UCA had stepped in at a number of key points throughout the process of human development.

These thoughts were more complex than the set-and-forget scenario of an unknown agency building a planetary regulator that was, in effect, an incubator for life. The idea that some entity, probably an advanced species from another galaxy, established a mechanism to foster life and then moved on seemed reasonable in the face of such evidence. But to have an agency that has maintained a periodic involvement with humankind across several billions of years is much harder to reconcile.

However, we decided to stick to our methodology of viewing the reasons 'why' ahead of dealing with the reasons 'why not'. We needed to review the material from ancient history and prehistory that had brought us to look so hard at the Moon.

Firstly we had to remember that it was because the stone structures from the fourth millennium BC were apparently created to study the Moon, that Alexander Thom began his lifetime quest to investigate them. Could these large standing stones be pointing deliberately at the Moon and the Sun? The orientation of the Megalithic structures certainly led him to identify the Megalithic Yard as being a unit of 82.96656cm – give or take 0.61cm. And this in turn led us to the findings laid out in this book.

As stated earlier, we had discovered that the Megalithic Yard was merely the starting point of a holistic measuring system that dealt with linear distance, mass, volume and time. It was an utterly brilliant system and we found that many modern units had descended from it, such as the imperial pint and the pound. We had been unable to imagine how the pound and the pint could have survived across so many millennia but it is a fact of mathematics that they are directly related, either by design or by an incredible series of coincidences.

To recap, the most intriguing fact about the Megalithic Yard lay in the way it had been ingeniously devised to fit accurately into the circumference of the Earth. Megalithic geometry was slightly different to the 360-degree geometry invented by the Sumerians, which is still in use more than four thousand years later. It had been based on 366 degrees, apparently (and very logically) because the Earth revolves once on its axis whilst it travels in its great orbital circle around the Sun. Under this Stone-Age system of geometry, each of the 366 degrees was split into sixty minutes of arc and each minute of arc into six seconds of arc.

The incredible beauty of the system is that when the globe of the Earth was treated as a huge circle, the polar circumference of the Earth is exactly the right size to give 366 Megalithic Yards to a polar second of arc.

We had been very surprised at the way the Megalithic Yard bisected the circumference of the Earth, but what we didn't expect to discover was any direct connection between the Megalithic Yard and other bodies within our solar system. And there are none – apart from the Moon and the Sun.

The Moon has a beautifully neat 100 Megalithic Yards to each second of arc, which could be a very odd coincidence if it were not for all of the other facts we discovered which point to a whole range of round numbers. And of course the Sun has an incredibly round

40,000 Megalithic Yards to each Megalithic second of arc. What a perfect way to announce an awareness that the Moon is exactly 400 times smaller than the Sun.

We also noted that whilst the Sun has 40,000 Megalithic Yards to a Megalithic second of arc, the metric system was designed so that the Earth's polar circumference would be 40,000 kilometres.

It had struck us as quite amazing that anyone more than 5,000 years ago could have created a unit of measure that worked as a perfect integer of the planet within such an elegant system of geometry – starting and finishing with the Earth's PIN number of 366. Whilst this was impressive, we were perplexed at the apparent impossibility of creating a unit and a geometry that produced beautifully round integers on the Earth, Moon and Sun. To do so should be as close to impossible as anything can get.

Units that are integer, within the same geometry, for two heavenly bodies would be very difficult – but three? That's ridiculous! And yet the sums spoke for themselves. The fact that the approach did not work for any other body in the solar system pointed to a very special relationship for the Earth, Moon and Sun.

The apparent impossibility of the Neolithic inhabitants having had the skills to develop such a marvellous system is now resolved when we introduce the unknown creative agency because, if it started with knowledge of the dimensions of the two original bodies (the Sun and the Earth), it could have engineered the Moon to made it fit the same rules. Our hypothesis was, therefore, to assume that the UCA somehow instructed the Stone-Age builders to adopt the system we call Megalithic geometry.

In our previous book, *Civilization One,* we speculated that the earliest records of the Sumerians and the ancient Egyptians were actually correct when they claimed that their own civilizations had been instructed in the arts and sciences by an external agency. In these records there are references to people called 'the watchers' who taught geometry, mathematics, astronomy, agriculture and other sciences. The indigenous population did not know where these people had come from and they described them as having superhuman powers, although they were clearly human beings and not gods.

In around 3100BC, ancient Egypt became a united kingdom and its period of recorded history began. At the same time, the Sumerians were building their great cities and developing sophisticated techniques of metalworking, glass manufacture and agriculture. In the Indus valley of the Indian subcontinent, the Harappa and Mohenjodaro civilizations were also constructing huge cities and in the British Isles, superb megalithic structures like Newgrange, Maes Howe and the Ring of Brodgar were being built. Is it not very strange indeed, that within such a precise period of time the whole world suddenly decided to step up a gear and enter into a period of true civilization?

We found it more than odd that these unconnected peoples should all take such a large step forward at exactly the same time. And we have recently come across very new information that made our suspicions even greater. On December 23rd 2004, new findings were published that markedly revise the dating of the first American civilizations. It reported that evidence now shows that the oldest civilization in the Americas dates back far earlier than previously thought – in fact right back to 3100BC, at which time complex societies and communal building suddenly appeared in Peru. This emerging culture was the first in the Americas to develop centralized decision-making, formalized religion, social hierarchies and a mixed economy based on agriculture and fishing.

One member of the team that has reported these findings in the pre-eminent scientific journal, *Nature,* is Jonathan Haas of the Department of Anthropology at the Field Museum in Chicago. He said:

> 'The scale and sophistication of these sites is unheard of anywhere in the New World at this time, and at almost any time.

These dates push back the origins of civilization in the Americas to something more parallel to those of the other great early civilizations.'[26]

Some of the settlements that are believed to have had at least 3,000 inhabitants included platform mounds, thought to be pyramids, central plazas, temples and housing. The largest pyramid at Caral, known as the Primade Mayor, is contemporary with the earlier Egyptian pyramids, dating from 2627BC. From this data, the archaeologists have concluded that there was large-scale communal construction and population concentration across the entire area.

Dr José Oliver, a lecturer in Latin American archaeology at the Institute of Archaeology at University College London, said: 'This confirms that by 3100BC monumental buildings were already under way, not just at an isolated site but across a whole region.'

As we have already stated, science is about recognizing patterns. Humans have not changed physically or intellectually over the last hundred thousand years but suddenly, just over 5,000 years ago, unconnected people around the world began building major structures and cities; but apart from some Sumerian–Egyptian interaction, these groups appear to have developed quite independently. Archaeology has not found obvious cross-cultural artefacts so it is assumed that they all blossomed at the same time through sheer coincidence.

But if they appeared worldwide because they had all benefited from the instruction of an unknown creative agency, one shouldn't necessarily expect an exact commonality of interpretation of these ideas. Nevertheless, it is clear that there are some significant cultural connections such as the building of pyramids and Venus worship.

There is, it seems, some very powerful, albeit circumstantial, evidence for an intervention by a highly advanced group more than 5,000 years ago. We have to admit, however, that we cannot conceive how any agency could have maintained contact with the Earth's development over several billions of years. Nevertheless, we do not see it as our place to reject information just because we cannot explain it. Everything depends on the ground rules of the observer: if someone refuses to look at obvious patterns because they consider a pattern should not be there, then they will see nothing but the reflection of their own prejudices.

Reciprocal Numbers

As we reflected on what we had found, the number play involved in the Earth–Moon–Sun system was nothing less than staggering. We were amused by the charm of this virtual machine especially when using the metric system. We looked at this little equation using kilometres:

$$\frac{\text{Moon x Earth}}{100} = \text{Sun}$$

This means that if we multiply the circumference of the Moon by that of the Earth, the result is 436,669,140km. If we then divide this figure by 100 we arrive at 436,669km, which is the circumference of the Sun, correct to 99.9 per cent.

How weird!

Of course, if we divide the circumference of the Sun by that of the Moon and multiply by 100 we get the polar circumference of the Earth. And, as we have pointed out, if we divide the size of the Sun by the size of the Earth and multiply by 100 we get the size of the Moon.

None of this is magic or pointless numerology. It may well be nothing more than an amusing coincidence but, given all of the ratio patterning we have observed, it would be foolish to ignore it.

However, the idea that kilometres can be meaningful to issues regarding the Moon is hard to swallow. Any reader could be forgiven for doubting what they read here. Nevertheless if anyone chooses to check out the numbers – it all works. And if you are still not sure about the idea, have a look at this fact; it certainly astounded us when we came across it.

The Moon has a sidereal rotation period of 655.728 hours, which means it rotates once every 27.322 Earth days. Given that the Moon has an equatorial circumference of 10,920.8 kilometres, this means that the Moon is turning at 400 kilometres per Earth day!

Just consider these unquestionable facts as a whole:

The Moon is one 400th the size of the Sun.

The Moon is 400 times closer to the Earth than the Sun.

The Moon is rotating at a rate of 400km per Earth day.

Coincidence? Well, maybe – or maybe not.

The Earth is rotating at 40,000 kilometres a day and the Moon is turning at a rather precise 100 times less. The Moon always faces the Earth as it travels on its orbit around our planet and yet the average distance is such that the equatorial rotational speed is precisely one per cent of an Earth day. These figures are entirely checkable and indisputable. How could all this be accidental?

Surely, only a fool would not wish to examine this situation further. Yet we have to be realistic about how some people will view our decision to consider the apparently impossible. We are well aware that many, and possibly most, experts will turn a blind eye.

Terence Kealey, aclinical biochemist and the Vice-Chancellor of the University of Buckingham, wrote an article in the (London) *Times* on November 15th 2004 under the title 'Who says science is about facts? They only get in the way of a good theory'. In this he recollected as follows:

'When Charles Moore was editing *The Spectator* he once asked me why, of his contributors, it was those trained in science who were the least honest ... Charles Moore had supposed that scientists would revere facts, but that supposition is a myth: scientists actually treat facts the way barristers treat hostile witnesses – with suspicion.

The mythmaker was Karl Popper. Popper was not a scientist but a political philosopher who proposed that science works by 'falsifiability': scientists discover facts; they create a theory to explain them; and the theory is accepted until it is falsified by the discovery of incompatible facts that then inspire a new theory ... Yet it is a myth that working scientists always respect falsifiability. Scientists often ignore inconvenient findings.'

We could not agree more, and therefore we will not be surprised if people ignore the possibility that the metric system just might be (crazy though it sounds) fundamental in some way to the Sun and the Moon as well as to the Earth. The fact remains that, for some reason, the kilometre demonstrates the essence of the Sun–Moon–Earth relationship, both in terms of size and orbital characteristics.

As if all of this isn't incredible enough we must also address the fact that the Moon has an orbit that makes it a 'mirror of time'. As we observed earlier, the Moon mimics the Sun at key points in the year. For example, whilst the Sun sets in the north at the time of the summer solstice, the Moon sets in the south and when the Sun sets in the south at the time of the winter solstice, the Moon unerringly sets in the north. This is an aspect of Sun and Moon associations that undoubtedly seemed like magic to our ancient ancestors and is yet another reflection of the current position and orbital characteristics of the Moon.

The Reasons Why Not

We have a constructed a scenario that fits all the facts but has deliberately ignored some of the challenging consequences that have arisen. We now need to deal with the reasons why this scenario might be wrong. Without the intellectual tether of having to conform to ideas that are within the bounds of what is already accepted, we have argued that an intelligent agency constructed the Moon to enable life to develop upon the planet we call Earth. We have taken a holistic view and we have not ignored any facts that we do not wish to have in our picture of what might have been.

The first problem that we thought we confronted – that of motivation, has been potentially answered in that it might be part of a grand quest to convert the Universe into an intelligent, self-aware single entity at the end of time. Such an idea would certainly seem to sit well with the principles of some Eastern belief systems such as Hinduism.

The Moon was already outrageously impossible before we introduced the issues of the intricate web of interrelated values, which we have argued is a deliberate message. With the number values that exist in the ratios alone, we fail to understand how anyone could seriously claim that they are coincidences. But the biggest challenge we have to confront is the issue of how the Megalithic Yard and the metric system came to be involved with an artificial Moon constructed as a life-support system for the Earth.

We cannot hide from the problem that, if our deductions are accurate, our unidentified creative agency has had contact with us at least once over the last 6,000 years. If this agency wished humans to know what they had done – and they (or it) are capable of making contact so recently – then why don't they just turn up right now and tell us what was done in the distant past, instead of leaving messages on the Moon?

We were puzzled. This did not seem to make sense.

As we debated this tricky point, we considered an alternative scenario that would not require direct contact from the UCA. Perhaps, we mused, the rise of the Megalithic system and even the metric system were programmed into our planet, to the extent that humans respond to these values quite naturally and without knowing why. Perhaps the gravitational effects of the Sun and the Moon interact with the Earth's own gravity and the effects of its spinning journey through space. It is known that the spinning orbit of the Earth does cause a disturbance in time-space, so maybe the value that we have called the PIN number, the value 366, is actually the heartbeat of our planet. Perhaps we cannot help but follow certain numerical patterns?

We were raising questions faster than we were solving problems but there was a strong logic to this notion. We knew that the ancient Sumerians had used a system virtually identical to the metric system in the middle of the third millennium BC, with a double kush that was 99.88 per cent of a metre. This unit was accompanied by others that were virtually a litre and a kilo.

We had already noted that the second of time appeared to be real in some way, rather than just an abstract convention. On Earth a pendulum that swings at a rate of once a second will have a length of a metre, with tiny variations dependent on the user's precise distance from the planet's core.

Perhaps the values programmed into the Earth by the UCA were so fundamental that any intelligent life form evolving on the Earth would respond to them. The relatively recent discovery that pendulums appear to go haywire during a total eclipse could point to brief interruptions of this Earthly harmony. We were aware that we were putting speculation upon speculation but it made sense. And we have to remember that we are not trying to displace any well-reasoned theory already in existence, so these possibilities have the benefit of being alone in fitting all of the known facts.

The bottom line to all this is that some unknown creative agency made the Moon out of parts of the Earth so that it would act as an incubator for life. The next question to confront was this: What was put into the incubator so that it would eventually grow into an intelligent life form? Setting up the hardware was impressive enough but what software was used?

CHAPTER TEN

The Impossible Accident

'A super-intelligence is the only good explanation' for the origin of life and the complexity of nature.
Professor Anthony Flew, December 2004

Not very long ago, religion was the only guide to the way the world was perceived. For right or for wrong the various scriptures of theological tradition provided a way of making sense of everything from the miracle of birth to the movement of the stars in the sky. But today we have rational thinking – we have science.

The word 'science' is from the Latin *scire,* meaning 'to know' and it is concerned with the organization of objectively verifiable sense experience. In other words, it makes sense of the way we see the world in a testable and verifiable way. It seems that there is nothing that science cannot explain given enough time and study. From Anthropology to Zoology, the people of the twenty-first century have experts who can explain where almost everything came from and how it works.

But science does have its limits. The Heisenberg uncertainty principle, for example, means that we cannot exactly know the position and the momentum of a particle simultaneously. We can choose one or the other – but we cannot have both. And there is at least one subject that science appears to be unable to explain. The origin of life.

In his book, *How to Think Straight,* Professor Anthony Flew has pointed out that practical reasoning and clear thinking are essential for everyone who wants to make proper sense of the information we receive each day. He stresses the importance of being able to quickly know the difference between valid and invalid arguments, the contradictory versus the contrary, vagueness and ambiguity, contradiction and self-contradiction, the truthful and the fallacious. These, he says, are the qualities that separate clear thinkers from the crowd.[27] After sixty-six years as a leading champion of atheism and logical thinking, Professor Anthony Flew has made sense of new information which has led him to state that science appears to have proven the existence of God. Flew's reason for this monumental about turn is the discovery of evidence that shows that some sort of intelligence must have created the world we inhabit. He has particularly pointed to the investigation of DNA by biologists, which has shown that an unbelievable complexity of the arrangements are needed to produce life;

leading to the conclusion that intelligence must have been involved.

We have bemoaned the lack of objectivity that often pervades the academic community but we must applaud a man who is prepared, at the age of eighty-one, to throw away the cornerstone of his life's work. That takes guts!

The first the world knew of Flew's change of heart was his letter to the August–September 2004 issue of the *Philosophy Now* journal where he stated: 'It has become inordinately difficult even to begin to think about constructing a naturalistic theory of the evolution of that first reproducing organism.'

Flew is a man of principle and when he was asked if his startlingly new ideas would upset some people, he responded by saying, 'That's too bad ... my whole life has been guided by the principle of Plato's Socrates: Follow the evidence, wherever it leads.'

How refreshing. That is exactly what we are trying to do with the information we have gathered about the Moon and the origins of life on our apparently designed incubator.

We have arrived at a point where we need to try and understand the emergence of life. And we find, at this precise moment, that the old assumptions about how life originated have been thrown out of the window.

The first question we asked ourselves is: What do we mean by 'life'?

We use the term to cover all kinds of organisms from cyanobacteria to plants and animals. The essence of life is reproduction, the formation of identical or near identical copies of a complex structure from simple starting materials. The increase of complexity involved in the formation of living organisms from their precursors distinguishes the processes of biological growth and reproduction from physical processes such as crystallization. This local increase of complexity can also be described as a decrease of entropy, which we have already speculated might be the motivation of the unknown creative agency that seeded and promoted life on Earth.

But where is the boundary of what is and what is not a life form. Is, for example, a virus a living entity? The standard answer is 'no it is not', but that is now seen as a very debatable point. Viruses cannot replicate on their own but can do so when they occupy a host. In the late nineteenth century, researchers realized that some diseases were caused by biological objects that were then thought to be the simplest and smallest of all living, gene-bearing life-forms. Throughout most of the twentieth century, though, viruses have been designated as non-living material.

All living organisms possess a genome, which is the set of instructions for making the body, and this is always composed of nucleic acid. It is usually DNA (deoxyribonucleic acid) or in the case of some viruses, RNA (ribonucleic acid). The genome consists of a number of genes, each of which is a segment of nucleic acid coding for a particular type of protein molecule. In October 2004, French researchers announced findings that blurred the boundary once again. Didier Raoult and his colleagues at the University of the Mediterranean in Marseille announced that they had sequenced the genome of the largest known virus, Mimivirus, which had been discovered in 1992. This virus, about the size of a small bacterium, contained numerous genes previously thought to only exist in cellular organisms. The virus is therefore a very smart bit of 'dead' matter or it is part of a unique club of entities only known to exist upon Earth.

The remarkable nature of living matter caused astrobiologist Paul Davies to observe in December 2004:

'Most people take the existence of life for granted, but to a physicist like me it seems astounding. How do stupid atoms do such clever things? Physicists normally think of matter in terms of inert, clod-like particles jostling each other, so the elaborate organization of the living cell appears little short of miraculous. Evidently, living organisms represent a state of matter in a class apart from the rest.'

The Solution that Identified the Problem

Back in 1953, when Watson and Crick discovered the helical structure of the DNA molecule and the general way that it coded the formation and replication of proteins in cells, it seemed that a plausible scientific explanation for the origin of life was about to be assembled. The laboratory synthesis of amino acids from basic chemicals further heightened the expectations that humankind was on the verge of creating a living cell.

It was suggested that the early Earth, through a mixture of volcanic activity and landmass weathering, had acquired oceans rich in nutrients and chemicals – known as 'the primeval soup'. It was in the constant mixing and intermixing of chemicals, and probably with the aid of lightning strikes, that the first primitive life had come into existence – or so the evolutionists suggested. Experts remained confident that the primeval soup theory was the most likely explanation and were convinced that, given time, someone would manage to create life in a laboratory.

Soon after Watson and Crick's discovery, Stanley Miller, a graduate student from Chicago University, co-operated with Harold Urey, a Nobel Prize winner, to recreate the exact circumstances that are believed to have existed in the primeval soup of the infant Earth. Their soup contained water vapour, hydrogen, methane and ammonia. It was estimated that lightning had played a part in the emergence of life, so Miller and Urey provided an electrical spark to their chemical soup and eventually succeeded in creating simple amino acids. 'Hooray!', they and everyone else concerned said, because amino acids are a major component of organic life. Unfortunately, more than half a century later, no one has come any closer to actually creating life than this.

It has also been pointed out that the amino acids created by Miller, Urey and others are but a tiny few of the constituents required for life. In any case, the experiment was very selective in its methods. Amino acids are referred to as being left- and right-handed, both of which were present in Miller and Urey's soup, whereas life uses only left-handed amino acids. What is more, the very electrical spark that created the amino acids would also have destroyed them, so they had to be artificially isolated in the experiment.

It might be thought reasonable that if life once formed in the oceans, it would continue to do so today. In reality this can't happen because the mixture of temperatures, chemicals and gases present is wrong. It was generally accepted that life could not spontaneously appear in an oxygen rich atmosphere and so the evolutionists had to suggest a very different sort of atmosphere on the infant Earth. (Oxygen, whilst preserving life, destroys organic molecules that are not alive.)

Generating life in the laboratory proved to be utterly impossible and researchers began to realize that new natural laws would need to be discovered to explain how the high degree of order and specificity of even a single cell could be generated by random, natural processes.

The DNA molecule is in the form of a double helix – rather like a ladder twisted into a spiral. The bases of the DNA are found in pairs and these make up the rungs of the ladder that carry the information to replicate the entity. When DNA copies itself, the ladder breaks down the middle of the rungs. New bases are matched to the bases of each upright and so the original DNA molecule then becomes two new identical molecules of DNA. Information necessary to build new proteins, and to perform other necessary chemical changes, is taken to various parts of a cell by another molecule, this one being ribonucleic acid (RNA). RNA is similar to DNA but is only a single helix. RNA is therefore the 'messenger' that allows the information held within DNA to be distributed and acted upon.

An important question remains, and it is one that science still cannot answer. How did DNA come about in the first place, because as things stand now, only DNA can create DNA.

Some chromosomes contain extremely long strings of DNA of more than a metre in length, which is colossal considering the microscopic nature of the DNA molecule itself. However, the question that has puzzled everyone concerned is the origin of this process, because all enzymes are proteins and protein synthesis must be directed by DNA. Yet, DNA replication cannot take place without these proteins. So, what came first – the protein or the DNA?

The problem goes right back to the origin of all life. But it is a problem that appears to have no answer. What is certain is that amino acids, nucleotides, lipids and other multi-atom molecules can be manufactured at random by heat, for example from lightning strikes. They can also come about from sunlight and other sources of energy that don't themselves have life. Many ideas have been put forward to explain the occurrence of DNA but none of them can be more than educated guesses.

But as we were researching this book a new theory appeared, and it is one that has gained favour with many experts. This theory suggests that DNA exists thanks to the presence of Earth's Moon!

Four billion years ago, the orbit of the Moon was much closer to the Earth than it is today. At this time, the Earth was spinning much faster on its axis and phenomenal tides were being raised on the Earth, by the constant passing of the Moon. With the Moon so much closer to

Earth the height of the tides would have been colossal (see chapter 5).

Richard Lathe, a molecular biologist at Pieta Research in Edinburgh, has suggested that within the primordial oceans, constantly dragged back and forth by the passing of the Moon, DNA could have been rapidly multiplied.[28]

One of the most commonly held theories regarding the origin of DNA is that it emerged when smaller, precursor molecules in the waters of the early oceans – 'primeval soup' – came together or were 'polymerized' into long strands. These long strands, it is suggested, became the templates for more molecules to attach themselves along the templates, which eventually resulted in double-stranded molecules like DNA.

Richard Lathe suggests that the problem lies in the need for some mechanism that would constantly break apart the double strands, in order to keep the process going. He maintains it would have taken some external force to dissociate the two strands.

It is at around 50°C that single DNA strands act as templates for synthesizing complementary strands, whereas at the higher temperature of about 100°C, these double strands break apart and this doubles the number of molecules. When the temperature falls, the process begins again. The number of replications grows exponentially with just forty cycles producing a trillion identical copies.

A billion years after the Moon came to orbit the Earth, it was extremely close to its host planet and the Earth was spinning much faster than it is now. The tides, as Lathe suggests, must have extended several hundred kilometres inland, which meant that coastal areas were subjected to rapid changes in salinity and this would have led to repeated and very frequent association and dissociation of double-stranded molecules similar to those of DNA.

As the huge tides advanced, salt concentrations would have been very low. Even modern double-stranded DNA breaks apart under such conditions, because electrically charged phosphate groups on each strand repel each other. However, when the tides receded, precursor molecules and precipitated salt would have been present in high concentrations. Lathe claims that this would have encouraged DNA-like double-stranded molecules to form, because high salt concentrations neutralize DNA's phosphate charges and this allows strands to stick together.

It is these constant salty cycles and changes in temperature that, Lathe says, would have amplified molecules such as DNA but he points out that the tidal forces were absolutely vital in the process. Whilst it is true that the Sun also creates tides on the Earth, these are of a very low magnitude compared to those caused by the much closer Moon. Three billion years ago it was closer still.

Without DNA there could be no life because it stands at the very heart of the replication of living matter. From the single-celled amoeba to the largest blue whale on our planet, DNA is the vital component that began life and which keeps it going. Perhaps Richard Lathe is correct and it was the presence of so large a Moon that began the chemical process that led to us, but it does remain a fact that despite all the theories, no scientist has yet managed to take the various chemicals that comprise life and arrange them in such a way that they become even the very simplest life form.

Lathe's theory could explain how the Moon caused the early replication of DNA but its origin remains a complete mystery, and many scientists are quite unsettled about the theory of how life came into existence in the first place. For example, David A Kaufmann PhD, of the University of Florida said, 'Evolution lacks a scientifically acceptable explanation of the source of the precisely planned codes within cells without which there can be no specific proteins and hence, no life.'

Admittedly David Kaufmann is a creationist, so maybe we can expect him to come to this conclusion. But then there is Professor Hubert P Yockey, a physicist from the University of California – who is most definitely not an adherent of creation and is concerned that discredited ideas continue to clog up the process of seeking out the truth. He wrote:

'Although at the beginning the paradigm was worth consideration, now the entire effort in the primeval soup paradigm is self-deception on the ideology of its champions...

The history of science shows that a paradigm, once it has achieved the status of acceptance (and is incorporated in textbooks) and regardless of its failures, is declared invalid only when a new paradigm is available to replace it. Nevertheless, in order to make progress in science, it is necessary to clear the decks, so to speak, of failed paradigms. This must be done even if this leaves the decks entirely clear and no paradigms survive. It is a characteristic of the true believer in religion, philosophy and ideology that he must have a set of beliefs, come what may (Hoffer, 1951). Belief in a primeval soup on the grounds that no other paradigm is available is an example of the logical fallacy of the false alternative. In science it is a virtue to acknowledge ignorance. This has been universally the case in the history of science as Kuhn (1970) has discussed in detail. There is no reason that this should be different in the research on the origin of life.'[29]

Yockey makes this statement because, like many other scientists, he cannot believe that the question regarding the emergence of life can be answered at all well by the primeval soup theory. Like the Double Whack theory of the Moon's birth – it is simply wrong and obfuscating progress to a workable explanation.

The main reason there is so much unrest about this question is because DNA cannot exist without life, and life cannot exist without DNA. The two are totally interdependent and create a chicken-and-egg situation that seems impossible to resolve.

It occurs to us that even the theories of Richard Lathe, on the way the Moon may have contributed to the rapid spreading of life through huge tides and chemical mixing, come no closer to explaining how life actually came about.

Some experts still claim that it must have happened by accident, presumably because the other possibilities are too hard to swallow. However, it would be far more sensible to claim that fairies from Neverland did it.

The Probability Problem

Nobody doubts that the information contained in a single gene must be at least as great as the enzyme it controls. However, just one average protein contains over 300 amino acids. In order to create the protein it would take a gene of DNA that would have to contain 1,000 nucleotides in its chain. Every DNA chain contains four sorts of nucleotide. This seems complicated but it results in a possible 4×10^{1000} possible forms. For those who do not realize, 4×10^{1000} represents the number 4 followed by 1,000 zeros.

These are values beyond all comprehension. To get some perspective on this, it is interesting to note that it is estimated that there are only 10×10^{80} particles in the whole Universe. One begins to realize how utterly impossible it would have been for complex DNA to be accidentally created in the primeval soup of the young Earth.

In the world of probability, some things are very likely to happen, others might sometimes happen but some can never happen at all. An expert in probability, Emile Borel (1871–1956) claimed that phenomena with very small probability don't occur. He estimated that there would be about one chance in 10×10^{50} for a small probability. Minute though these odds were, they weren't remote enough for more modern experts in probability. William M Dembski, associate research professor in the conceptual foundations of science at Baylor University and a senior fellow with Discovery Institute's Center for Science and Culture in Seattle, decided to go further. He estimated that there were 10×10^{80} particles in the Universe and wondered how many times per second an event might occur. The number he came up with was 10×10^{45}. He then calculated the number of seconds from the beginning of the Universe to the present time and then, to make sure he was erring on the side of caution, he multiplied this number by one billion and arrived at the number 10×10^{25} seconds. He now multiplied all the figures together achieving a result of 10×10^{150} for his Law of Small Probability.[30]

For a minimum living cell there are 60,000 proteins of 150 configurations.[31] Joseph A Mastropaolo, an expert who has tackled this problem at length, estimates that the probability of the evolution of this first cell would be an absolutely staggering 1 in $10 \times 10^{4,478,296}$ or 10 followed by 4,478,296 zeros. This exceeds Dembski's estimation for Small Probability by such a great margin that were it not for the fact that DNA does clearly exist, no self-respecting scientist could uphold the possibility of it having originated by chance.

If every particle in the Universe had one chance for every second since the beginning of time – we still would not have DNA.

In case there are readers who doubt Mastropaolo's scepticism regarding the possibility of DNA creating itself from scratch, it is interesting to see that he is far from alone. Peter T Mora of Macromolecular Biology Section, Immunology Program, National Cancer Institute, Bethesda, Maryland wrote: 'The presence of a living unit is exactly opposite to what we would expect on the basis of pure statistical and probability considerations.'[32]

The English scientist J D Bernal said, way back in 1965: 'The answer would seem to me, combined with the knowledge that life is actually there, to lead to the conclusion that some sequences other than chance occurrences must have led to the appearance of life as we know it.'[33]

And to add to the list of dissenters regarding a theory that clearly doesn't hold water, primeval or not, we have the opinion of the late Professor Sir Fred Hoyle, one of the most respected astronomers who has ever lived. 'Rather than accept that fantastically small probability of life having arisen through the blind forces of nature, it seemed better to suppose that the origin of life was a deliberate intellectual act. By "better" I mean less likely to be wrong.'[34]

However, no matter how great and how many the howls of indignation at this complete disregard of probability, one of the fundamental tools of science, it remains a fact that DNA did occur somehow. As the saying goes, nature abhors a vacuum of any sort. No matter how much Professor Yockey may suggest that if we have no viable theory we should exist without it until one is discovered, it seems that to many scientists a twisted and broken paradigm is better than none at all.

After all, the alternative might be unthinkable to most experts. We might, for example, have to consider the possibility of a 'mind' behind the creation of DNA, even if we can accept evolution as a viable theory once DNA existed.

The majority of scientists would prefer to break their own rules rather than to evoke the deity, but even Professor Sir Fred Hoyle was left with the only conclusion that could occur to him, namely that the Universe was under some sort of 'intelligent cosmic control'.[35] Is this the way forward? If we are going to be truly honest, bearing in mind the utter impossibility of the chance occurrence of DNA, might we have to accept that 'God spoke and it was so'?

Who can blame Anthony Flew for turning a lifetime's work on its head and saying: 'A super-intelligence is the only good explanation for the origin of life and the complexity of nature.'

However, Flew's definition of God bears little resemblance to the deity of Judeo-Christian-Islamic tradition, which he describes as being depicted as 'omnipotent Oriental despots – cosmic Saddam Husseins'. He is actually describing something as open as our own 'Unknown Creative Agency' – which presumably might mean virtually anything from a sublime single entity to a galactic federation of planet seeders!

The Seeds of Life

Some sixty years ago, when quantum theory first emerged, physicists thought the mystery of life was about to be resolved. By looking at the tiniest building blocks of matter it was starting to explain how everything worked – so it surely would also explain the essence that we call life. They were to be disappointed, but recent developments have raised the hopes of some scientists that the nature of first life might be explained by new levels of understanding about sub atomic behaviour in biology.

In 2004, these new ideas caused NASA to convene a workshop of leading scientists to discuss the subject of 'quantum life' at their astrobiology laboratory in Ames, California, where discussion covered fields such as nanotechnology and quantum computation.

Nanotechnology is concerned with the manufacture of artefacts or machines that are assembled on an atom-by-atom basis. A nanometre is an almost unbelievably small unit of length. A human hair is typically about one 10,000th of a metre in diameter and a common cold virus is approximately one thousandth of this size. A typical protein unit making up the coating of such a virus is typically ten nanometres thick – equivalent to about 100 atomic diameters, or the size of one of the amino acid groups making up that protein molecule.

A whole new world of technology is envisaged from building self-replicating machines that could, for instance, carry out surgery at a cellular level inside the human body. However, an increasing number of scientists are suggesting that nature may have used this idea a long time ago. As Professor Paul Davies has pointed out, the living cell is full of nanomachines designed and refined by biological evolution. And he posed the question: 'Could it be that some of them acquired their amazing properties by deploying fancy quantum tricks?'[36] He says: 'One vital part of a cell's reproductive machinery is a little motor, called a polymerase enzyme, which crawls along unzipped strands of DNA and forges the links that match up the unpaired nucleotide bases with complementary bases floating through its environment.'

Apoorva Patel of the Indian Institute of Science, believes that living cells may use quantum mechanics to boost their information-processing efficiency, which could explain why the genetic code is the way it is, and why it is found in all organisms. As Davies points out, quantum theory describes atoms and molecules as waves, which can overlap and combine coherently – known as superposition. This means that the normal rules of time/space do not apply and an atom can exist in a superposition of excited and unexcited states, or of states corresponding to several spatial locations at the same time. These superpositions are expected to be the basis of quantum computers that will be able to hunt for a target among a jumble of data. This is said to be equivalent to finding a name in a telephone directory when you know only the phone number.

The role of quantum theory in the origin of life is not yet clear. But it seems that the new technologies that humankind is now investigating may be at the root of life itself. Paul Davies acknowledges that life somehow emerged from the ferment of the quantum molecular world, and he adds:

'The role of quantum processes in living matter is still unclear. It is entirely possible that quantum mechanics was the midwife of life, but has played an insignificant role since ... All scientists agree that life somehow emerged from the ferment of the quantum molecular world. The key issue is on which side of the quantum-classical divide the transition to life occurred. Niels Bohr once said that anyone who is not shocked by quantum mechanics hasn't understood it. I believe that anyone who is not shocked by life hasn't understood it. The question before us is whether quantum mechanics is shocking enough to explain life.'

It seems to us that whoever seeded life on our planet, all those billions of years ago, was using a form of self-replicating 'technology' which will eventually come to be understood. And we might not be that far from that understanding now.

Who Built the Moon?

At this point we are as sure as we possibly can be that the following statements are true:

- The Moon is approximately 4.6 billion years old.
- The Earth is approximately 4.6 billion years old.
- The Moon was manufactured from lighter materials taken from the young Earth.
- The Moon was made as an incubator to foster life on Earth.
- The manufacturer of the Moon seeded life on Earth.
- Evolution as described by Darwin is broadly accurate.

The manufacturer of the Moon left deliberate messages, intended to be read by humans at this point in geological time, to draw attention to what they have done.

It appears reasonable to assume that the manufacturer of the Moon (the UCA) has a good reason for wanting humans to understand what was done. The UCA could have seeded life and moved on, if its motive was pure altruism. It therefore seems that it is important to work out who built the Moon.

It seems certain that we have only identified the first 'introduction' aspect of the message from the UCA. The details of the message are likely to hold the key to the next phase of human development: information that will change our destiny forever.

We believe that we have succeeded in identifying the key numbers that will be used in deeper layers of the detailed communication. We trust that others will take up the challenge of interpreting other aspects of the message, but our immediate task is to try and work out who built the Moon.

And we believe there are only three possibilities.

CHAPTER ELEVEN

Childhood's End

'And God made two great lights; the greater light to rule the day, and the lesser light to rule the night: he made the stars also.'
Genesis 1:16, The King James Bible

Good things and terrible things have always happened to mortal man. The warmth of spring, the survival of infants, the provision of animals to hunt, plants to harvest and freedom from disease must surely be the work of an unseen force with powers far beyond than that of mere people. So too, the ills and woes of failed crops, floods and death wrought upon whole tribes by war and desperate want. It must be the will of the gods.

Thank the gods, fear the gods, appease the gods.

Religion is as old as the stories that humans first told. From the early Stone Age to the Internet Age, humankind appears to need the power of deities that inhabit an unseen world and yet have the power to affect the lives we live. The greatest love and the greatest hate spill forth in the name of gods.

Today, the great religions of the world tend to describe the gods in the singular as God, even though they all refer to many aspects under different names.

The Hindu tradition has ideas that are increasingly seen as corresponding with modern science. It perceives the existence of the cyclical nature of the Universe and everything within it, where the cosmos follows one cycle within a framework of larger cycles. The Universe has been created and reached an end, but it represents only one turn in the perpetual 'wheel of time', which revolves infinitely through successive cycles of creation and destruction. This cycle of creation and destruction of the Universe could be seen as a series of Big Bangs and Big Crunches, where all matter explodes outwards from nowhere and then recedes back again. Within these gigantic cycles the soul also undergoes its own cycle, called samsara – where death and rebirth sees the same souls repeatedly reincarnated.

Meanwhile, Christianity is a broad church indeed, covering an incredible span of beliefs. At one end of the spectrum there are many scientific thinkers – including at least two Fellows of the Royal Society. One of them, John Polkinghorne, was a mathematical physicist before resigning his position as a professor at Cambridge University in 1979 to be ordained as priest in the Church of England. Polkinghorne has since devoted his life to exploring the connections between science and theology, describing the Universe as open and flexible – a place where patterns seem to exist and where he says the 'providential aspect' cannot be ruled out.

Many Christians fully support science and have no problem with evolution, quantum mechanics or the big bang origin of the Universe. For them it is simply a question of the authorship of the blueprint that obviously exists. The designer of all this is their God. And yet they also believe in an event that others would find incredible. Without wishing to be disrespectful, we would précis that event as follows: The initial intellect that created everything became a man and died, nailed to a wooden post, some two thousand years ago, before briefly returning to human life and then transferring back to His ethereal state somewhere outside of the physical world. This anthropoid interlude for this creator deity (many billions of years after the start of the Universe) is believed to compensate for the bad behaviour of those people who accept this story as real, thereby ensuring a pleasant continuation of consciousness after their physical body has ceased to be alive.

At the other end of the Christian belief are the creationists. They hold that a collection of ancient Canaanite and Mesopotamian myths, from at least three separate traditions and first written down in the sixth century BC by Jewish priests, are a literal account of how the world came into being. They take an uncomplicated view of life and consider that all species are unchanging and derive their forms from an unchanging, divine blueprint. To a creationist a rose is a rose is a rose, and it is foolish to think that a rose bush could become a daffodil, or an apple tree. They see God's plan as timeless and unchanging, with separate types of plants and animals that have nothing to connect them. For them the world and everything within it was created in six days of a single week, somewhere around 4004BC.

It is of central importance to creationists that there is an absolute divide between humans and other animals. They often use the phrase 'don't let them make a monkey out of you' in the mistaken belief that evolutionists claim that humans developed from monkeys.

Buddhist philosophy is evolutionary and in many ways agrees with mainstream scientists. Buddha taught that all things are impermanent, constantly arising, becoming, changing and fading. Buddhist philosophers consequently rejected the Platonic idea of production from 'ideal forms' as being the fallacy of 'production from inherently existent other'. According to most schools of Buddhism there is nothing whatsoever that is inherently or independently existent.

Buddhist philosophers have always accepted that the Universe is billions of years old and they have no corresponding creation myth to that of the Judaeo-Christian tradition. Unlike creationists, Buddhists believe that both humans and animals possess sentient minds that survive death.

There are many people today who are agnostic, meaning that they do not see any proof of God but neither do they believe it to be impossible that there could be a God. Perhaps a small minority of the world are true atheists believing that all matter, including their own self-awareness, is merely the culmination of multiple accidents occurring at random within the basic laws of physics.

The classical argument for God has been that there must have been a 'first cause' but this is considered to be invalid by relatively modern philosophers such as David Hume and Immanuel Kant, because the thesis is negated by its own premise. If everything must have a first cause then what made God? It therefore follows that the Universe could arrive spontaneously just as much as God could.

But, it occurs to us, what if God and the essence of the Universe were, and are, the same thing?

God in Contact

Human societies have probably always developed the idea that the world they see around them must have a conscious mind behind it. And the Judaeo-Christian tradition holds that God has had quite regular contact, particularly with His chosen people, from Adam through characters such as Enoch, Noah, Abraham, Moses, Ezekiel, Isaiah and John the Baptist (Jesus Christ cannot count because that would be God talking to Himself). Following the crucifixion of Jesus, God, or members of His ethereal team, are believed to have had contact with inspired individuals from St Paul to Joan of Arc, and there have been many miraculous appearances at locations such as Lourdes in France, Fatima in Portugal and Knock in Ireland.

These visitations are held to be wonderful by believers and considered baloney by others. Apart from the apparent miracle of existence itself, all other aspects of God require faith. Faith could be described as intellectual belief that transcends normal standards of proof. In other words, the individual with faith holds things to be true that are not evidenced in a form that rational science would accept.

But what would happen if God suddenly turned up in an unambiguous way; if the creator of the Universe appeared, in person, on the Earth with positive proof of identity?

Logic says it could not happen because it is likely that only the agnostics would be happy. Those who would be most likely to welcome the coming of God are, by definition, the people with the most complex belief systems. And every group (possibly except one) would be disappointed. Would Mormons be told they had it right after all or maybe Roman Catholics?; or maybe some followers of Mohammad or Siddhartha Gautama or any one of the countless prophets down the ages.

Imagine the Pope and the Dalai Lama sitting shaking their heads in disbelief as it turns out that the Australian Aboriginal people and those of the Japanese Shinto faith both had it right when they called God Izanagi. Surely, it would have to be those with the most religion who would have the most to lose.

But then, it is not likely that they are all correct in some way and that God is actually non-denominational. What if He now considers that the childhood of the human race is over and we are now grown up enough to be told the true mysteries of existence – he might choose to make gentle contact to let us know that in some way 'we had arrived'.

It is our initial thought that the number patterns built into the Moon and its relationship with Earth, could be a first global contact with God Himself. Such an event would change everything. If God formally made his presence known, who would dare wage a war in His name? The world might listen carefully instead of proclaiming its right to speak on His behalf from the churches, synagogues, mosques and temples around the globe.

What evidence is there that this message could be from God?

The first problem is one of definition. What do we mean when we speak of God? For recent convert, Anthony Flew, God is simply the creative force that does not interact with people, but for many millions of others He is a benign father figure who listens to their prayers.

Upon reflection, the only way to deal with this point is to ignore it. If the human species has reached the end of its 'childhood', the nature of God will be appreciated in a new light anyway.

The most fundamental case for the God scenario, when it comes to the message the Moon has to impart to us about its artificial construction, is that any entity who created our world is God, almost by definition. Scriptures from all around the world attribute the making of our planet and the heavens to a creative force that usually has a special relationship with humankind. That relationship is so special in Christianity that it is central to the very belief system that the creator of our world actually became a man for thirty-three years some two millennia ago.

The fact that the numbers used in the Moon's message are in base ten, implies that the UCA knew that the intelligent species that would evolve on Earth would have ten fingers. God would know that. It is also clearly the case that the UCA knew that it would be at this particular point in the Earth's history that humans would be ready for the next stage of their relationship with God.

The story told in the Book of Genesis in the Old Testament would turn out to be remarkably correct and even the Christian creationists would be right in part.

> 'In the beginning God created the heaven and the earth.'

In this scenario, God did create the Earth and the heavens, and by regulating its attitude with the Moon caused it to have liquid water on its surface:

> 'And the earth was without form, and void; and darkness was upon the face of the deep. And the Spirit of God moved upon the face of the waters.'

In its early years the Moon was orbiting close to the Earth, gradually slowing down both its own spin and producing a spin that gave regular days and nights:

> 'And God said, Let there be light: and there was light. And God saw the light, that it was good: and God divided the light from the darkness. And God called the light Day, and the darkness he called Night. And the evening and the morning were the first day.

And let them be for lights in the firmament of the heaven to give light upon the earth: and it was so. And God made two great lights; the greater light to rule the day, and the lesser light to rule the night: he made the stars also.'

The tilt of the Earth was held steady by the Moon and the Earth enjoyed regular days, years and changing seasons:

'And God said, Let there be lights in the firmament of the heaven to divide the day from the night; and let them be for signs, and for seasons, and for days, and years.'

The early Moon was huge and powerful as it orbited close to the Earth raising colossal tides every time it passed overhead. If the Moon had not been created, the seas of the Earth would cover virtually all of the planet leaving little dry land:

'And God said, Let there be a firmament in the midst of the waters, and let it divide the waters from the waters. And God made the firmament, and divided the waters which were under the firmament from the waters which were above the firmament: and it was so.'

Thanks to its close proximity, the Moon's tidal surges travelled far inland, constantly stirring the life-nurturing soup of the oceans, ready for the moment life arrived. As more advanced life developed, plant life came first:

'And God said, Let the earth bring forth grass, the herb yielding seed, and the fruit tree yielding fruit after his kind, whose seed is in itself, upon the earth: and it was so. And the earth brought forth grass, and herb yielding seed after his kind, and the tree yielding fruit, whose seed was in itself, after his kind: and God saw that it was good.'

The first animal life began in the oceans before spreading to land and into the air:

'And God said, Let the waters bring forth abundantly the moving creature that hath life, and fowl that may fly above the earth in the open firmament of heaven. And God created great whales, and every living creature that moveth, which the waters brought forth abundantly, after their kind, and every winged fowl after his kind: and God saw that it was good.

And God blessed them, saying, Be fruitful, and multiply, and fill the waters in the seas, and let fowl multiply in the earth. And God said, Let the earth bring forth the living creature after his kind, cattle, and creeping thing, and beast of the earth after his kind: and it was so. And God made the beast of the earth after his kind, and cattle after their kind, and every thing that creepeth upon the earth after his kind: and God saw that it was good.'

Millions and millions of creatures came and went, slowly changing into more complex life forms and eventually gaining intelligence and self-awareness. One branch of mammals climbed into the trees and later returned to the plains as hominids – our ancient, ape-like ancestors. There were many species of hominid that learned to use primitive tools and that survived as hunter-gatherers. As recently as 25,000 years ago there were still three species of human: *Homo floresienis, Homo neanderthalis* and *Homo sapiens.* The Neanderthals had larger brains than ours and we can be sure that they laughed and talked and cried – their burial practices even suggest that they may have had religious belief. But today, we are alone:

'And God said, Let us make man in our image, after our likeness: and let them have dominion over the fish of the sea, and over the fowl of the air, and over the cattle, and over all the earth, and over every creeping thing that creepeth upon the earth. So God created man in his own image, in the image of God created he him; male and female created he them.

And God blessed them, and God said unto them, Be fruitful, and multiply, and replenish the earth, and subdue it: and have dominion over the fish of the sea, and over the fowl of the air, and over every living thing that moveth upon the earth.'

The period during which we have learned to walk upright and have developed such large brains that our Mothers risk their lives in giving birth to us, is miniscule in terms of the period the Earth has existed. It has taken us only a couple of million years in total. The amount of time we have been bright enough to look with knowing eyes at our world has been much less than that, merely a few tens of thousands of years. We learned how to hunt and to survive from the bounty of nature and eventually we became farmers, living fixed lives and establishing villages that became towns and eventually cities.

Maybe six or seven thousand years ago something remarkable may have happened. Whoever or whatever had manufactured the Moon returned. In an operation that possibly involved a whole series of 'visits', the cipher necessary to crack the code of the message, that had been so carefully encapsulated into the Moon, was given to humanity. This 'key' was the Megalithic system of measurement and geometry and specifically the Megalithic Yard. The Moon's creator must have been aware that if the Megalithic Yard was written into the stone circles and avenues of what is now Britain and France, someone would eventually recover the information and rebuild the entire system in all its splendour.

This was clearly not enough. Another series of visitations took place, not long after the first but this time to another proto-civilization far from the first, between the rivers Tigress and Euphrates, in what is today the area known as

Iraq. Here a second system of mathematics and geometry was seeded, this one less related to the mathematical certainties of the Earth and its relationship with the Moon but more closely tied to everyday life. It was the forerunner of much that was to follow and when the rise of science came along, humanity invented the metric system, which almost eerily reflected what the Sumerians had been so carefully taught. The astronomer priests of Sumer were shown that the whole world, its size, mass and volume, could be derived from the most humble source possible – a single seed of barley. (See Appendix Five.) This plant had clearly been genetically engineered not only to be of fantastic use to humanity but also to lock into the dimensions and mass of the Earth in an almost unbelievable way.

Mythology and folklore tells us time and again that 'messengers' were sent in the remote past to teach humanity the rudiments of civilization and we now know why. None of this is beyond the capabilities of God and it is likely that a percentage of readers will already be convinced that this must be the solution to the message contained in the Moon.

God could quite easily have created the Moon and done so well within the laws of physics He had ordained. It would have been His deliberate intention that the life He seeded on the young Earth would eventually give birth to a thinking, rational species that was, in some way, made in His own image. His interest in humanity, when it eventually evolved, remained as He had quite clearly intended. We can see a situation in which the Deity sent messengers to lay the foundations of an eventual recognition of the message which would lead to the first tangible proof of the existence of a Creator.

Nothing is beyond the mind or capability of God. We have endowed Him with unparalleled power and timelessness. But for countless generations the reality of God has resided in 'faith' rather than 'proof'. Perhaps those with religion will resent the suggestion that God has removed the need for faith.

The humorous and thought-provoking writer, the late Douglas Adams, played with this notion in his book *The Hitchhiker's Guide to the Galaxy.*[37] Adams created a remarkable creature known as the Babel Fish, that when placed in anyone's ear could act as an intergalactic speech translator. So remarkable was the existence of this little fish that people said it must stand as irrevocable proof of the existence of God because nothing so amazing could possibly come about by chance. However, it was pointed out that since God existed by faith alone – and not by proof – the absolute proof of his existence meant for certain that he could not exist.

'I never thought of that,' said God, and disappeared in a puff of logic.'

It is clear that if we accept that God was responsible for creating the Moon and that He specifically incorporated within it proof of what He had done, we must begin to look at Him in a very different light. In a world in which religion has been diminishing in importance, and particularly in the technological West, an acceptance of God's direct intervention in our part of the solar system might see thousands or millions of people flocking back to Church. The most fervent creationists may abandon their insistence that the Earth is only a few thousand years old and might accept that God did indeed work his magic through evolution. The recognition of God's special pact with life, and especially with humanity, might fund a push towards ecumenicism and a coming together of the world's fractious religions.

Unfortunately it is equally likely that the reverse would happen because power-bases, religious or secular, have always shown a reluctance to diminish in importance. Clearly, if we are looking at God's true covenant with humanity through his intentional creation of the Moon, with its attendant and obviously deliberate messages, no existing belief pattern can be any more important than another and the whole basis of religious dogma is in doubt.

We could not criticize anyone who wishes to attribute the message to God. But neither could we argue with anyone who says that God does not need to leave messages coded into ancient stone circles that He already knows will eventually be recognized by humanity. If we are ultimately left in no doubt as to his existence, the whole procedure has been somewhat unnecessary. God is capable of showing Himself to humanity at any time He chooses, with absolutely no ambiguity or the remotest uncertainty.

Everything about the Moon and its addition to the solar system seems to speak of a message that 'must' be imparted one day and of a series of deliberate 'humanlike' interventions that would ensure this was the case.

Further to this, we might argue that the Moon was almost certainly added to our part of the solar system as an afterthought. It had to be, because the very material from which it was made came from the already existent Earth. God could quite easily have made the Earth a haven for life in its own right. It has to be remembered that it was the 'shortcomings' of the Earth that necessitated the addition of the Moon to the planetary system. Surely the God of the human imagination is all-powerful and has no shortcomings.

We cannot deny that a world in which humanity was certain of the existence of God, and in which there was no longer any doubt about what He represented, 'might' become a more cohesive and peaceful place and we did not turn away from this possibility lightly. However, we have tried to approach our research from a genuinely scientific point of view (we would argue that our approach is more scientific and less based on enshrined belief than that of many so-called scientists.) This being said, we felt ourselves obliged to look at other possible solutions to the questions raised by the evidence we had amassed. Those who wish to attribute the creation of the Earth–Moon system to God will continue do so, though we felt it impossible to stop searching. We are cognisant that by His sheer timeless power God can be used as a cure-all to answer any question. That has been the pattern of humanity across the ages and it is not one we feel constrained to follow.

In short, there are other possibilities that might prove to be just as surprising but considerably more plausible.

Postscript to this chapter

This chapter was completed during the closing days of 2004. On the morning of Sunday December 26th an Earthquake five miles beneath the ocean floor, west of Sumatra, produced a tsunami with the power of more than 10,000 atomic bombs. Travelling at speeds of up to 800 kilometres an hour it tore into coastal areas all around the Indian Ocean causing devastation that was as sudden as it was terrible. Many tens of thousands of people died within minutes and millions more were left to grieve for their lost loved ones and to struggle against hunger, thirst and the threat of consequential disease.

The event was so powerful that the entire Earth moved.

Geologist Kerry Sieh of the California Institute of Technology said 'It caused the planet to wobble a little bit.' As the Indian Ocean's heavy tectonic plate lurched underneath the Indonesian plate there was a shift of mass towards the planet's centre, causing the globe to rotate faster and shortening the period of our planet's rotation by some three microseconds. A team of researchers at NASA's Jet Propulsion Laboratory in Pasadena, California also found that the incident caused the Earth's tilt to be increased by 2.5cm.

The mobility of the Earth's crust was central to the emergence of life and now the residual shifting of tectonic plates causes death and destruction to those too near to the event. If the careful design of the Earth and its Moon were the work of God, His life-bringing mechanisms are, in this instance at least, working against the interests of His chosen species.

The events in the Indian Ocean horrified the world. In Britain the Archbishop of Canterbury, who leads the Church of England, was deeply troubled. Dr Rowan Williams, writing eight days later in the *Sunday Telegraph* questioned the nature of God's interaction with humans:

'The question: "How can you believe in a God who permits suffering on this scale?" is therefore very much around at the moment, and it would be surprising if it weren't – indeed, it would be wrong if it weren't. The traditional answers will get us only so far. God, we are told, is not a puppet-master in regard either to human actions or to the processes of the world. If we are to exist in an environment where we can live lives of productive work and consistent understanding – human lives as we know them – the world has to have a regular order and pattern of its own. Effects follow causes in a way that we can chart, and so can make some attempt at coping with. So there is something odd about expecting that God will constantly step in if things are getting dangerous. How dangerous do they have to be? How many deaths would be acceptable?

So why do religious believers pray for God's help or healing? They ask for God's action to come in to a situation and change it, yes; but if they are honest, they don't see prayer as a plea for magical solutions that will make the world totally safe for them and others.

All this is fair enough, perhaps true as far as it goes. But it doesn't go very far in helping us, one week on, with the intolerable grief and devastation in front of us. If some religious genius did come up with an explanation of exactly why all these deaths made sense, would we feel happier or safer or more confident in God? Wouldn't we feel something of a chill at the prospect of a God who deliberately plans a programme that involves a certain level of casualties?'

If a single entity that we could reasonably call God did indeed establish the Earth and its Moon so that we might evolve, He might be obliged to work within His own rules of the Universe. Creating a life-bearing planet required a ploughing of the surface and this is a process that cannot be switched on and off like a light switch. Dr Williams presumably has a problem because he believes in a God who is in on-going contact – a God who can choose to respond to individual prayers. But maybe the situation is not like that.

The title we chose for this chapter is 'Childhood's End'. This seemed to be a fitting summation for the discussion of the possibility that God had made the world and had, from the outset, built into it a message that we would understand when we were sufficiently emotionally and intellectually mature. We were aware that Arthur C Clarke had written a novel with this title more than half a century earlier with a very different but not unconnected theme.

Sir Arthur is an inspired writer and his ideas expressed in *2001: A Space Odyssey,* have been discussed in this book. When we realized that the Indian Ocean tsunami had caused a massive loss of life in Sri Lanka, we were concerned for him, because we were aware that he is wheelchair bound in his home near the beach in Colombo. Thankfully Sir Arthur was not hurt and was able to write an account of what had happened in his adopted country.

He wrote: 'I have no idea if God had any scenario in mind when this happened. In a way, the disaster was a random event, but at the same time nothing in this world is totally random, there is always cause and effect.'

All of this could very much describe a God who has a working plan that appears to be less than perfect. Tectonic plates were necessary to create us but their current movements are simply a small, incidental effect of a far greater cause. Are we to believe that in the mind of God the ultimate end justifies the sometimes very painful means?

CHAPTER TWELVE

Extra Terrestrials

'...it's entirely possible, in my view, that we could retrieve a message from another world within just a few decades...'
Seth Shostak – Senior Astronomer, SETI

The idea that intelligent creatures might exist somewhere else in the cosmos has fascinated humanity ever since the invention of the telescope revealed that our world is but one amongst countless others. At first some people wondered if there were people living around the supposed seas on the Moon and others feared invasion from near neighbours, particularly Mars.

In 1858 an Italian astronomer called Secchi announced that he had seen *'canali'* on the surface of Mars, and in 1877 Giovanni Virginio Schiaparelli, an astronomer at the Milan Observatory, produced drawings of these features. Though the most accurate translation of the Italian word *'canali'* would have been 'channels', it was translated into English as 'canals'. With the completion of the Suez Canal fresh in people's minds, the interpretation was taken to mean that huge artificial waterways had been discovered – which amounted to evidence of intelligent life.

Debate raged over the findings, with Schiaparelli himself stating that there was no reason to suppose that the canals were artificial. The discovery sparked the imagination of a young man named Percival Lowell who was at the beginning of what was to be a distinguished career in astronomy. He was one of the first to realize that it was far more sensible to site observatories in out-of-the-way places, such as deserts or on mountaintops, where smoke and light spillage from cities would not diminish the astronomers view of the heavens. He was the driving force behind the creation of the Lowell Observatory in Flagstaff, Arizona, in 1894.

Professor Lowell studied linear features on Mars with his twenty-four-inch telescope and developed theories about the habitability of Mars, based on his estimate that the planet had an average temperature of 48°F. The Lowell Observatory made consistent observations of the Martian canals and Lowell personally maintained that the linear features were indeed of artificial origin.

When spacecraft reached Mars, scientists expected to discover what the canals really were but they found that there were no canals and almost no straight lines on the planet at all. We have to conclude that either the Martians have camouflaged them rather well over the last century or, infinitely more likely, a generation of astronomers were imagining things at the limits of their optical telescopes.

The idea that there could be real Martians was a popular worry that was brilliantly used as the plot by H G Wells in his novel *War of the Worlds.*

A wave of mass hysteria gripped thousands of radio listeners in October 1938, when a dramatization of this book was broadcast and led unsuspecting listeners to believe that an interplanetary conflict had started, with invading Martians spreading death and destruction across New Jersey and New York.

The next day the *New York Times* reported on the scare:

'A weather report was given, prosaically. An announcer remarked that the program would be continued from a hotel, with dance music. For a few moments a dance program was given in the usual manner. Then there was a "break-in" with a "flash" about a professor at an observatory noting a series of gas explosions on the planet Mars.

News bulletins and scene broadcasts followed, reporting, with the technique in which the radio had reported actual events, the landing of a "meteor" near Princeton N.J., "killing" 1,500 persons, the discovery that the "meteor" was a "metal cylinder" containing strange creatures from Mars armed with "death rays" to open hostilities against the inhabitants of the earth.'

By far the majority of experts now accept that if advanced life of any sort does exist in places other than the Earth, we will almost certainly have to look towards interstellar space in order to find it. But our greater knowledge of outer space has not quelled the public's appetite for close-encounter stories.

The famous Roswell incident is believed by many to be an extraterrestrial encounter. It is said that a UFO crashed in the New Mexico desert in July 1947 and the debris was removed to an army base in Fort Worth, Texas.

A US government cover-up is said to have tried to pass off the event by stating that the debris was actually part of a radar unit from a weatherballoon.

Rumours about the existence of secret alien bases located in various places, such as the Moon, under the ocean, or in a tropical rain forest have persisted. Some people have gone so far as to claim that they have worked on secret UFO projects for the government and seen UFOs at military installations.

According to a recent poll, some three million Americans believe that they have encountered bright lights and incurred strange bodily marks indicative of a possible encounter with aliens. Psychological tests confirm that these 'abductees' are rarely psychotic or mentally ill in any usual sense of the term.

It makes us wonder whether humans are simply prone to having some kind of neural dysfunction involving optical illusions. Maybe the decline of old-style belief in mythical creatures like fairies and goblins and in religious imagery such as angels or the Virgin Mary, has caused people to have new kinds of hallucinations. Where people once thought they saw the 'little people' dancing in a circle of light or a heavenly messenger with a glowing halo, the bright lights in their heads are now translated as alien contact.

Whilst the debate continues about everything from Roswell to crop circles, it has to be admitted that there has never been any proof of alien contact – and it is, of course, impossible to prove the negative. However, the probability of contact does seem extremely small, given the vast amounts of space and time involved.

The solar system, of which the Earth forms a small part, is only one of many even in our own corner of our galaxy – the Milky Way. Astronomers have identified stars that definitely have planets orbiting them, so the state of affairs within our own solar system is certainly not unique. An interesting finding has been that larger, gaseous planets in other star systems, much like Jupiter and Saturn in our own, have been discovered to have an orbit that is always very close to their host star. From these early indications it seems that our planetary arrangement is unique, which just might not be accidental.

It is a fact that if Jupiter were not just over five times more distant from the Sun than we are, advanced life on Earth would not exist. This giant planet is positioned as a 'catcher' of space objects that would otherwise impact into the Earth. A dramatic example of this was seen in July 1994, when twenty-one fragments of the comet Shoemaker-Levy 9 smashed into Jupiter at speeds of up to half a million kilometres and hour, creating fireballs larger than the planet Earth.

If we are right about the Moon being constructed to act as an incubator, the manufacturer would have been pleased to note that Jupiter and Saturn were in very unusual and perhaps unique outer orbits. If it were not so, they would have to have caused them to be in this position – which would suggest that the entire solar system could have been designed for the benefit of humankind!

Whether or not our solar system is a happy accident, it is estimated that there are a thousand million other stars in our galaxy alone, any one of which could possess a planetary system where life might have evolved and even flourished. Beyond our galaxy there must be stars with Earth-like planets beyond counting. Bearing these facts in mind, it surely appears unreasonable to believe that only our tiny little green planet is alone in producing a self-aware species.

However as we have previously noticed, setting out to actually meet our intergalactic or extragalactic cousins seems hopeless, even if we knew where they were located. But this may not be the end of the story.

Time is not a fixed concept. If a person could travel close to the speed of light, they would experience a severe slowdown in time, relative to a slower moving object. At light speed, time stops completely, relative to something moving at a much lesser speed. Because of this 'time stop', a photon that travels at the speed of light would not experience distance and time in the normal way. So from the photon's point of view, it could go from one end of the Universe to the other instantly, while from an outside point of view it would take about thirteen billion years.

Still stranger, scientists have found the need to speculate about the existence of a particle called a 'tachyon' that can travel faster than light. But theoretically at least, travelling faster than light would result in an individual going backwards in time. So the tachyon is something of a mystery at the moment, with scientists having to calculate the activity of these particles with time working in reverse.

So, just maybe, there will be ways to work around the problem of travelling at speeds close to, or even above, the speed of light.

Next, there is the possibility of intergalactic communication using what physicists call 'quantum entanglement', that can happen to sub atomic particles. If quarks with identical spin are paired and separated, and the spin of one is changed, the other changes its spin instantaneously to match that of its partner – no matter how far apart they are separated. Einstein called this phenomenon 'spooky distance', and it suggests that some force, not yet understood, must be capable of travelling in folded space in some manner or may not exist at all in space as we know it, and therefore not be restricted to the effects of travel.

It is therefore not inconceivable that other advanced creatures have found a way to bridge the chasm of space-time between their planet and ours. But we are not able to deal with such technology yet, even though we can envisage its existence. Right now, as far as we know, we cannot greet them face to face, but as we pointed out in Chapter Eight it might be possible to listen to them or even talk to them.

As we have also noted, recent publications by leading academics such as Paul Davies, Christopher Rose and Gregory Wright, are suggesting that physical artefacts are a far better way of communicating across the vastness of space. Paul Davies has stated that a far more reliable way for any alien species to contact us would be to leave artefacts in the vicinity of planets likely to spawn intelligent life that, given sufficient advancement on the part of such a developing species, it could not fail to recognize.

And so, the question that confronts us is: Could aliens have built the Moon from the very substance of the Earth in order to allow our development, and then left a physical message of what they had done in the very dimensions and movements of the bodies?

We believe that the message we have detected in the Moon and its relationship to the Earth is so amazingly differentiated from the 'background noise' of all other measurements that it forms a break-through for humanity. Certainly, if a message of such clarity and consistency was received from beyond our planet by means of good old-fashioned electromagnetic radiation, the personnel at SETI would be jumping up and down with joy.

If the message from the UCA is attributable to aliens we have already speculated that its motive could simply be a desire to progressively transform the matter in the Universe from a chaotic condition to an ordered state of self-awareness. One can image that, given enough time, all of the matter in existence could be united in a single thinking entity. Astronomer Royal, Sir Fred Hoyle, wrote a novel called *The Black Cloud*[38] in which he speculated about a cloud of space matter that had such instantaneous interaction between its particles it was, effectively, a single living entity. Could this be the long-term goal for all intelligence? If so, we will need to understand what has happened in the case of our own planet much more clearly so that we will be able, in due course, to take part in this ultimate mission for the Universe.

If we accept alien intervention in our distant past, we have to ask how these visitors from elsewhere could have known that the fruits of their labours would come to have ten fingers and therefore work in base ten. A possible answer is that all successful life forms come to intellectual maturity with these characteristics, but the whole notion does seem odd.

Furthermore, there is the problem of how the alien Moon builders came to use Megalithic geometry and kilometres to incorporate elements of the message. This too seems unusual. What is more, as we have observed, there appear to have been visits to the Earth by the UCA in much more recent times. This would suggest that the alien visitors, having manufactured the Moon, would have had to return to the Earth over four billion years later in order to pass the Megalithic message onto the developing human culture in Britain and France. We find it difficult to imagine a culture or society that could endure for such a vast period of time. It is much more likely that such a civilization would have gone the way of inevitable evolution, managed somehow to destroy itself, or simply grown bored with the whole experiment in only a tiny fraction of the time involved.

If readers wish to believe that aliens are responsible for this message, we would have to say that this is a theory worthy of further investigation. For our part, we can see no direct proof that this has been the case, and there seem to be factors involved that make the alien hypothesis unlikely to be the answer we are seeking. However, there is a third, and altogether more amazing option yet to consider and it is one that appears to fit the bill in every respect.

CHAPTER THIRTEEN

The Möbius Principle

'Let us make man in our image, after our likeness'

God: Genesis 1:26

For those people who call themselves creationists, the Bible is the word of God. But which Bible is the authentic one? There are countless versions of the books contained in both the Old and New Testaments and the oldest versions have been carefully dissected to reveal the different styles of authorship woven into the fabric of the stories. Two of the three main traditions – the Yahwist and the Elohim (a word meaning gods in the plural) – talk of a specific sequence of creation. This deals with the arrival of plants, then good and evil, then animals and next women.

The third, priestly tradition has a sequence of creation that is rather more in line with modern theories about evolution. First comes light followed by heaven, the Earth (land and then sea), vegetation, then the Sun, Moon and stars. Next come birds and fishes and finally man and women together.

An interesting fact is that the first two traditions use the Hebrew *yàsar* for the creative act of making man, which has a simplistic or crude implication of being shaped, as a potter models clay objects. Both also use the word *demut* for likeness, which implies similarity or looking the same. However, in the Priestly tradition, (the version that has God talking to his wider council about making man in their image) He uses a very different word. In this case the word *bàrà* is used for the creation of man and this is a word that carries a more complex, creative value. Next we find *selem* as the chosen word for the use of the creator's image, which means something more like a precise duplicate. *Selem* is a term directly related to the Canaanite word for Venus that is associated with resurrection and therefore rebirth of the individual.[39]

We find it strange that a supposedly singular God is talking to others around him, even before humans have been created. He has already made the Sun, Moon and Earth and supplied the oceans along with plant and animal life – but to whom is he talking? And why do they all, whoever 'they' are (including God Himself) apparently have heads with noses, ears and eyes, bodies with arms and legs and presumably even genitalia?

Why is God, along with his undisclosed team, human in appearance?

It is not our place here to try and make sense of Judaeo-Christian myth, but we came to find the idea fascinating and surprisingly plausible. The Bible has been edited, changed and added to by a succession of people who wanted it to support whatever they deemed to be true. Early Christians even accused the Jews of having incorrect versions of their own scriptures when they were found to differ from the texts the early Christians had doctored. In terms of Christianity, it seems unlikely that a passage that involves God talking to others before he created humanity would have survived, had it not been for an important aspect of the new Christian faith. This was the 'new' concept of the trinity – where God is said to comprise three separate entities including his living human mode as Jesus.

We are not attempting to claim that the Bible provides us with any evidence for the authorship of the message we had discovered, but a close look at the situation did lead us to a tantalizing thought.

Could the only known intelligent life force in the Universe be responsible for the message? To be blunt: Could modern humans have built the Moon?

There is obviously one very substantial issue of logic to address here, which is obviously the time gap of 4.6 billion years between the creation of the Moon and the present era. Clearly, if humanity created the Moon, this would have to be explained. In reality, this may not be the obstacle it appears to be, because leading scientists are currently debating the possibility of travelling backwards in time. Virtually everyone speculating about time travel is agreed that the associated mathematics indicates it should be possible. We will come to the problem of travelling in time shortly, but for the moment let us put the issue of the time gap aside and consider the reasons why the Moon's message might be from closer to home than we ever dreamed could be possible.

The hypothesis we originally laid down was:

1   The Moon was engineered by an unknown agency circa 4.6 billion years ago to act as an incubator to promote intelligent life on Earth.

2   The unknown agency knew that humanoids would be the result of the evolutionary chain.

3   That unknown agency wanted the resulting humanoids to know what had been done and they left a message indicated by the dynamics of the Moon and its relationship with the Earth.

Firstly, it has to be acknowledged that there are no other possible candidates that we know of anywhere in the Universe. God exists by faith and not as a result of evidence, and aliens may or may not exist. It is entirely possible that we are totally alone, either in our part of space or in the whole of the Universe. In any case, who would have more to gain from a life-producing planet than the very intelligent creature that has most benefited from its existence, namely humanity?

The question of how the UCA could have known that the intelligent species on Earth would evolve with ten fingers and therefore adopt base-ten arithmetic, at a time when the Moon was exactly where it is today, is answered instantly if humanity is the agency we are seeking. The mystery simply dissolves if we are that unknown creative agency.

Another difficult issue to explain has been how the UCA could possibly have used Megalithic and metric units as part of the message. Once again, this scenario resolves the problem. Indeed, it adds to the message because it makes it very clear that the UCA 'has to be' humans from our future, travelling back in time to manufacture the Moon.

The motive for the message becomes obvious and absolutely necessary. If humans do not become alerted to the need to manufacture the Moon as an incubator for life – we would not be here.

However, there is the problem we can't avoid. Humanity might be described as having been reasonably technologically advanced for around 100 years. The Moon came into being some 4,600,000,000,000 years ago. We have to admit that this does represent a bit of a gap.

The answer can only be time travel.

Tomorrow's yesterday

Time is perceived as flowing like a river from the past into the future and we are all riding the wave in one direction. But what if it were possible to head back upstream? Not necessarily for humans themselves, though that cannot be ruled out, but for preprogrammed super-machines; equipment so sophisticated that it could engineer planetary-sized objects. After all, most spacecraft today are unmanned units that carry out all kinds of experiments, take photographs and even analyse samples of alien rock. It would not therefore be hard to imagine a project team from our relatively near future designing and deploying 'chronobots'[40] to construct key elements of the past.

But is time travel a dream or a possible reality?

For most people such thoughts cause headaches. The question that anyone will reasonably focus upon is: If humans went back in time to build the Moon so that there would be humans – where did humans come from?

It seems like an impossible loop – but is it stranger than the age-old conundrum about the chicken and the egg? Logically, it is necessary to have a chicken to lay an egg, yet one needs an egg for that chicken to have sprung from.

A creationist would have no problem as their God manufactured the first chicken with an ability to lay eggs. The evolutionists would be a little sneakier and say that a creature that was not quite a chicken laid an egg that produced a mutation that was the first proper chicken. So, the egg came first.

It really is not worth losing sleep about such problems, as the only way to deal with any paradox is to simply accept it.

Today, we are programmed with a need for neat, predictable Newtonian-style logic. Simple cause and effect – so that if 'A' happens 'B' will result. People everywhere seem willing to accept the idea that we were either created by God, or that we exist due to a mega-series of flabbergastingly beneficial accidents. Look at these two possibilities again and then ask yourself if it is any more far-fetched or unreasonable to suggest that, as a species, we went back to create our own life-giving planet system and ultimately ourselves? (For some reason, to the religiously-minded, the insurmountable question of 'Who made God?' can be safely ignored, as can the ridiculous improbability of an infinitely flowing stream of beneficial serendipity to non-believers).

Humans throughout history have generally had a psychological need for a higher authority, whether it be a supreme deity or the laws of physics. Thankfully, that is not necessarily the whole story at all.

The debate about time travel goes on amongst the experts as it has done for many decades. Generally speaking, philosophers don't care for the idea, for a whole host of logical or illogical reasons, though some of them are coming round in the face of the latest evidence. Meanwhile physicists are becoming increasingly certain that time travel is possible, and they have the mathematics to back up what is far from being a simple hunch.

Whilst the idea of travelling into the past is so counterintuitive for most people that they just cannot get their heads around it, a physics heavyweight and a philosophy heavyweight from Oxford University have another view. They once teamed up to confront the apparent paradox that seems to forbid the highly fluid present penetrating the apparently frozen structure of the past. David Deutsch and Michael Lockwood have the problem in context; saying about the quantum physics of time travel: 'Common sense may rule out such excursions – but the laws of physics do not.'[41]

Most people have a real problem with the idea of time travel, and the so-called 'grandfather paradox' encapsulates why the idea appears to assault common sense so strongly. The idea is that if a young man was able to travel back from the present time to, say, 1950, he might kill, or cause his grandfather to be killed before his own father was born. If this were to happen, it would mean that he could not exist and therefore could not have killed his grandfather. The problem just goes around in apparently impossible circles. The only solution appears, at first view, to be to consider all such journeys as utterly impossible – if for no other reason than to save us from terminal confusion!

However, Deutsch and Lockwood are not so easily fazed and they remain unconvinced about the need to protect our sensibilities from issues of reality just because laypeople tend to become confused. In an article published in *Scientific American* they discuss another apparent time paradox that deals with the possibility that even knowledge does not seem to require a beginning.

They refer to the grandfather-killing scenario as being an 'inconsistency paradox' and then they discuss another type of apparent time-traveller violation of logic that they call a 'knowledge paradox'. This is an apparent violation of the principle that knowledge can only come into existence as a result of problem-solving processes, such as biological evolution or human thought. In the example, they talk about a hypothetical art critic who goes back in time to visit a famous artist from the previous century who, the critic realizes, is only producing very mediocre work. The time traveller shows the painter a book containing reproductions of his later and greater works, which he then proceeds to carefully copy in every detail with oil paints onto canvas. This means that the reproductions in the book exist because they are copied from the paintings and the paintings exist because they were copied from the reproductions. So, where did the inspiration come from?

'This kind of puzzling paradox,' say Deutsch and Lockwood, 'once caused physicists to invoke a chronology principle that, by fiat alone, ruled out travel into the past.' But they believe that travelling into the past does not violate any principle of physics, however much it seems counterintuitive to the average person. Furthermore, the Oxford duo state that quantum-mechanical effects actually facilitate time travel rather than prevent it, as some scientists once argued.

They explain the basics of the concept of time by pointing to Einstein's special and general theories of relativity where three-dimensional space is combined with time to form four-dimensional space-time. Within this, everyone's life forms a kind of four-dimensional 'worm' in space-time, with the tip of the worm's tail corresponding to their birth and the top of the head to the person's death. The line along which the 'worm' lies is called the person's (or object's) 'worldline' and each moment of time is a cross section of that worldline.

Einstein's general theory of relativity predicts that massive bodies, such as stars and black holes, distort space-time and bend worldlines. This is believed to be the origin of gravity – and, for example, the Earth's worldline spirals around that of the Sun, which in turn spirals around that of the centre of our galaxy. Deutsch and Lockwood propose that if space-time becomes really distorted by gravity some worldlines would become closed loops where they would continue to conform to all the familiar properties of space and time in their own locality, yet they would become corridors to the past. They state:

> 'If we tried to follow such a Closed Timelike Curve (or CTC) exactly, all the way around, we would bump into our former selves and get pushed aside. But by following part of a CTC, we could return to the past and participate in events there. We could shake hands with our younger selves or, if the loop were large enough, visit our ancestors. To do this, we should either have to harness naturally occurring CTCs or create CTCs by distorting and tearing the fabric of space-time. So a time machine, rather than being a special kind of vehicle, would provide a route to the past, along which an ordinary vehicle, such as a spacecraft, could travel.'[42]

So, world-class physicists like Professor Deutsch can conceive of potentially giant spacecraft voyaging backwards in time. Perhaps such craft could be filled with chronobots that could even self-replicate to take on a task that might take hundreds of thousands, or even millions, of years. Building an object the size of the Moon with preprogrammed orbital requirements is unlikely to be a quick exercise. But time would literally be on their side.

There are various ideas about how these time-travel enabling CTCs might be formed. The mathematician Kurt Gödel found a solution to Einstein's equations that describes CTCs within a rotating Universe and they also appear in solutions of Einstein's equations describing rotating black holes. But there are many practical problems including the evidence that naturally occurring black holes are not spinning fast enough. Maybe a technique will one day be found to increase their rotation rate until safe CTCs appear.

The physicist John A Wheeler from Princeton University has famously suggested shortcuts through space-time that he calls 'wormholes', and other scientists have shown how two ends of a wormhole could be moved, so as to form a CTC.

Professor Deutsch has become a champion of the many-Universes theory, first put forward by Hugh Everett III in 1957, where everything that can happen does happen. For this reason, the supposed paradoxes of time travel simply do not exist. In the scenario where the man kills his grandfather, he does not exist in the one single Universe where the murder is committed, but he does in the ones where he fails to assassinate his forebear.

Deutsch and Lockwood conclude that there is no scientific objection to time travel, saying in their article:

> 'The idea that time travel paradoxes could be resolved by "parallel Universes" has been anticipated in science fiction and by some philosophers. What we have presented here is not so much a new resolution as a new way of arriving at it, by deducing it from existing physical theory ... These calculations definitively dispose of the inconsistency paradoxes, which turn out to be merely artifacts of an obsolete, classical worldview.'

They appear to be suggesting a loop in time that has a twist in it so that contact is made with a near identical parallel existence, through which the time traveller can arrive at a time and place that always has them present.

Their thought-provoking article concludes with the authors pointing out that science says time travel is theoretically possible. As a result, the onus is on those who wish to argue otherwise to prove their case:

> 'We conclude that if time travel is impossible, then the reason has yet to be discovered. We may or may not one day locate or create navigable CTCs. But if anything like the many-Universes picture is true – and in quantum cosmology and the

quantum theory of computation no viable alternative is known – then all the standard objections to time travel depend on false models of physical reality. So it is incumbent on anyone who still wants to reject the idea of time travel to come up with some new scientific or philosophical argument.'

And many experts agree. Physicist, Matt Visser of Victoria University of Wellington, has compiled a short list of the time travel opportunities that have turned up since Einstein showed us how to warp space-time. He has said that Einstein's general theory of relativity not only allows time machines to exist, it is 'completely infested with them'.

Others fear the concept of time travel, even though they have not been able to demonstrate that it cannot be done. 'I think most of us would like to get rid of time machines if we possibly could,' says Amanda Peet of the University of Toronto. 'They offend our fundamental sensibilities.'

The only argument that has been made against time travel comes from the famous Cambridge phycisist, Stephen Hawking, in the form of his 'chronology protection conjecture'. This suggestion boils down to the notion that the Universe might have a built-in time cop, so whenever anyone is on the verge of constructing a working time machine the time cop turns up and shuts the operation down before it has a chance to damage the past. However, there are no time cops evident in the laws of physics, so, at the moment, the chronology protection conjecture is simply wishful thinking.

As far as our scenario is concerned, humans exist because, at some future point, we will return to the time when our planet was a young lump of unstratified matter and then we shall make the Moon.

Once complete, our Moon worked its magic and life began, evolving eventually into an intelligent, ten-fingered species that uses Megalithic and metric units. The message had to be built into the very nature of the structure or else we would miss the cue to understand what we need to do.

But how can we do it and when will we do it?

Ronald Mallett, a Professor of Theoretical Physics at Connecticut University, already believes he has found a way to create a CTC or time machine using light. He has identified that a circulating beam of light, slowed right down to a snail's pace, may well be the key to the door of time travel because, although light has no mass it does bend space. The realization that time, as well as space, might be twisted by circulating light beams caused Mallett to team up with other scientists at Connecticut University in 2001, with the intention of building a prototype, saying, 'With this device time travel may become a practical possibility.'

Mallett decided that if he added a second light beam, circulating in the opposite direction to the first, it would increase the intensity of the light enough to cause space and time to swap roles. Inside the circulating light beam, time runs round and round, and, what to an outsider appears to be time becomes like an ordinary dimension of space. A person walking along in the right direction could actually be walking backwards in time – as measured outside the circle. So after walking for a while, you could leave the circle and meet yourself before you have entered it.

However, it turns out that the energy needed to twist time into a loop is enormous, and when Mallett reviewed his progress he realized that the effect of circulating light depends on its velocity: the slower the light, the stronger the distortion in space-time.

By strange good fortune, Lene Hau, a phycisist at Harvard University, has slowed light from the usual 300,000km per second to just a few metres per second, and almost frozen its progress completely. Mallett was ecstatic saying, 'The slow light opens up a domain we just haven't had before. All you need is to have the light circulate in one of these media.'

Maybe current scientists will crack the problem of time travel but it seems logical to expect the necessary instructions to be contained in the deeper layer of the Moon's message. However, it seems likely that black holes may be at the root of the technology required.

The black holes of deep space are the gravitational remains of dead stars. They are super-dense, bottomless pits in both space and time that are capable of sucking in almost infinite amounts of material, including light. Everything a black hole swallows gets compressed into an unimaginably tiny central region called a 'singularity' in which the atoms are crushed into an unmoving whole. If the Earth were to become as dense as a black hole, it would be smaller than a golf ball. (And they say you can't compress water!)

There seems to be no way to get any information about what is happening inside a black hole, as even light is trapped inside. However, Cambridge physicist Stephen Hawking proposed a way in which black holes do radiate matter and slowly dissipate until they eventually disappear in a final mega-burst of radiation.

Amazingly, scientists are becoming increasingly confident that they will be able to create black holes on demand using the new atom-smashers due to come on line in 2007. It is believed that the new Large Hadron Collider (LHC) being built astride the Franco-Swiss border west of Geneva by the European Centre for Nuclear Research (CERN) will be able to create black holes at the rate of one per second. The LHC is an accelerator which will bombard protons and antiprotons together, with such a force that the collision will create temperatures and energy densities not seen since the first trillionth of a second after the big bang. This should be enough to pop off numerous tiny black holes, with masses of just a few hundred protons. Black holes of this size will evaporate almost instantly, their existence detectable only by dying bursts of Hawking radiation.

This work is at an early stage but it may well prove to be the beginnings of a platform that could drive the search for the technology to enable time travel.

If humans from our future did travel to the distant past to create the incubator that would produce our own species, it does make complete sense of the message left to us. We have to imagine that our ability to complete such an awesome task must be hundreds or even thousands of years ahead of our current level of capabilities. However, what if the instructions of how to proceed were contained inside the message itself? If this was the case, the development time might be cut to a minimum.

Maybe a question we should be asking ourselves is why the message was so carefully timed to reveal itself at this particular time. Could it be that we have so far only seen what is little more than a 'waving flag' to alert us to a greater message that tells us exactly what must be done in order to fulfill our own destiny? Maybe the central pattern revealed by the mutual orbits of the Earth and its Moon and, quite separately, by the relative sizes of 366.3x27.3=10,000 is the most fundamental key of all.

At this stage there are two entirely separate questions that need to be answered:

1   To what are we to apply the cipher?
2   If humans created the Moon as an incubator for life, where did the seeds for germination come from?

The answer to both final elements of this ultimate riddle may well rest in the same place: DNA.

The secrets of the Genome

The Human Genome Project, completed in 2003, was a thirteen-year mission to unravel the secrets of the minute data store that carries all the information needed to make a human being, what we call DNA. The key goals of the project were to:

1   Identify all the genes in human DNA, of which there are believed to be approximately 20,000–25,000
2   Determine the sequences of the three billion chemical base pairs that make up human DNA

Professor Paul Davies has published an idea that strikes a real chord with the findings laid out in this book. He does not criticize the people from SETI for constantly sweeping the skies with radio telescopes, in the hope of stumbling across a signal from deep space, but he is realistic about the chances of success. He points out that it is inconceivable that aliens would beam signals at our planet continuously for untold aeons, merely in the hope that one day intelligent beings might evolve and decide to turn a radio telescope in their direction. And if the aliens only transmit messages sporadically, the chances of us tuning in at just the right time are infinitesimal.

However, he does not write off the idea of contact: 'But what if the truth isn't out there at all? What if it lies somewhere else? Now may be the time to try a radically different approach.'[43]

Davies uses the idea we have already reported of a 'set-and-forget' technique of communication, whereby the information content of the message may have to survive for hundreds of millions of years. He acknowledges that a conventional artefact placed on the Earth's surface would be almost certainly overlooked, even if it did somehow survive. He then suggests that an altogether better solution would be: '...a legion of small, cheap, self-repairing and self-replicating machines that can keep editing and copying information and perpetuate themselves over immense durations in the face of unforeseen environmental hazards.'

This sounds like pure science fiction but he continues by saying: '...Fortunately, such machines already exist. They are called living cells.'

What a brilliantly simple idea. We have already established that large sections of the scientific community are openly saying that DNA absolutely could not have spontaneously arrived – it must have been designed. So, why would the manufacturer not use it to contain a message?

Is it possible? Is there spare space in there for a message?

As Paul Davies confirms, the cells in our bodies, and anything else that lives for that matter, contain messages set out billions of years ago. He also says that the idea that aliens have deliberately hidden messages inside DNA has been 'swirling around' for a few years, and has been championed in recent times by the Apollo astronaut Rusty Schweickart. But, says Davies, on the face of it, there is a serious problem.

Living cells are not completely immune to change, and mutations introduce random errors that become stored as information, and, over a long enough time span, they would turn the original message into 'molecular gobbledygook'. Davies then reminds us that there is so-called 'junk' DNA: sections of the genome that seem to serve no useful purpose. These areas could be loaded with messages without affecting the performance of the cells and some parts of that junk DNA are in highly conserved regions that are therefore relatively safe from degradation.

When a team of genomic researchers at the Lawrence Berkeley National Laboratory in California presented their own findings in June 2004, the audience gasped in unison. Those listening, simply could not believe what they were hearing from Edward Rubin and team who were reporting that they had deleted huge sections of the genome of mice without it making any discernable difference to the animals. The result was truly amazing because the deleted sequences included what is known as 'conserved regions', which were previously assumed to have been protected because they contained vital information about functions.

To find out the function of some of these highly conserved nonprotein-coding regions in mammals, Rubin's team deleted two huge regions of DNA from mice, containing nearly 1,000 highly conserved sequences shared between humans and mice. One of the removed chunks was 1.6 million DNA-bases long and the other was over 800,000 bases long – which should have caused the mice to have serious problems.

All DNA can acquire random mutations, but if a mutation occurs in a region that has a key function, the individual will die before they are able to reproduce and therefore the damage to the information will be removed from the species. This protection mechanism means that the most vital sequences of DNA remain virtually unchanged – even between species. So by comparing the genomes of mice and men, geneticists had hoped to pick out those with the most important functions by studying the conserved regions.

The geneticists were utterly perplexed because the regions they removed made no difference to the mice in question, so there seemed to be no reason why these non-coding sequences, apparently functionless parts of the DNA, should be protected from change. Why should they matter? It is like having the world's finest encrypted security system built into your waste bin.

Any burglar who observed that your rubbish had so much apparently unnecessary protection would immediately suspect that you were hiding something of great value in an unexpected place. And that is the thought that occurred to Paul Davies. He believes there could be a message from extra terrestrials in what has been referred to as junk DNA.

We suspect he might be right about the message but not about the author. He says:

'Looking for messages in living cells has the virtue that DNA is being sequenced anyway. All it needs is a computer to search for suspicious-looking patterns. Long strings of the same nucleotides are an obvious attention-grabber. Peculiar numerical sequences like prime numbers would be a clincher and patterns that stand out even when partially degraded by mutational noise would make the most sense ... if a sequence of junk DNA bases were displayed as an array of pixels on a screen (with the colour depending on the base: blue for A, green for G, and so on...'

He then asks what the message could contain and notes that one segment of DNA, chopped out by Rubin and his team, contained more than a million base pairs – enough, he says, 'for a decent-sized novel or a potted history of the rise and fall of an alien civilization.'

And this would be from just one part of the junk DNA.

As we digested Davies' suggestion about number sequences making a screen we were immediately reminded of how the numbers that we have identified as the lead key of the message produce 10,000 – or if the decimal point is removed from the values we get the following:

$3663 \times 273 = 999{,}999$

As close to a million as makes no difference.

These are the PIN numbers of the Earth and Moon doubly cross-referenced by their orbital periods and relative masses. Without the decimal point, they describe a screen (possibly a computer monitor) that has a million pixels with sides of 3663 and 273.

One of the 'high security' sections of apparently empty genome removed by Rubin's team had just over a million elements. It would be more than interesting to apply the 3663x273 matrix to this data.

What will it tell us?

It may well give us vital information about building equipment that moves matter backwards in time and it will tell us where to start the process of planning to build a Moon! It is likely that it will also instruct us where to look for further information.

If we are correct, we are all carrying this 'treasure map' in our hearts, our brains and even our hair. But so too is every living creature on God's Earth.

'Let us make man in our image, after our likeness,' said God.

Could it really be true that a team of humans will control the creation of our world and seed it with DNA so that humans will evolve in our own image? Will a future president of the United States of America, or perhaps a Director General of the United Nations sanction the launch of a mission to create these mammoth, but necessary, changes to the past, whilst quoting the words from verse 26 of the first chapter of the Book of Genesis?

This is not a blasphemous thought. Some Christians and indeed people of other religions might find this idea offensive because it appears to suggest that we humans are God. But this is not the case. It merely suggests that we acted and will act on creative information that was originated somewhere else by some elemental force that transcends all Universes – all parallel realities.

The awe and mystery of existence remains intact and for those that want to call that essence 'God', He remains unchallenged.

However, the account we describe here does sit well with the scriptures of the great religions. Genesis is remarkably accurate and, as it turns out, the creationists may not be entirely wrong about a grand design in which humans were existent from the start. They will have to adjust their dating assumptions, which do not come from the Bible anyway. And they will have to accept that evolution was just a mechanism within the grand design.

The Hindu perception of the way the Universe works also remains intact, and the only adjustment they might adopt would be to accept that the cycle of rebirth has twisted into reverse as well as going forwards. We doubt the intellectuals within Buddhism will have a problem with this.

We see this process of self-conception as something akin to a Möbius strip, named after August Ferdinand Möbius, the nineteenth-century German mathematician and astronomer. Möbius was a pioneer in the field of topology. Along with his contemporaries, Riemann, Lobachevsky and Bolyai, Möbius created a non-Euclidean revolution in geometry.

The simple construct that is a Möbius strip can be made with a strip of paper by joining the ends with a 180° twist. It then only has one surface and one edge that goes around forever. Without the half-twist it would have been impossible to move from one side to the other without crossing an edge – but suddenly the barrier does not exist. If one travelled in a straight line on a Möbius surface one would return to the starting point.

Figure 13. The world's most famous graphic artist Maurits Cornelis Escher (1898–1972) was fascinated by the imagery of the Möbius surface.

We see an analogy with humankind who evolved from DNA, seeded on Earth some 3.5 billion years ago by ourselves, just a little in our future. When we reach the point of being able to travel back in time we will have completed a circuit of the single-sided loop and then move off into a new trajectory.

Once the idea of time travel is accepted as a scientific possibility, there is no problem with the idea that humans in the future engineered both DNA and the life-nurturing Earth–Moon system billions of years ago. We exist because the right circumstances were present for life to develop – and so why does it matter whether a super-entity (God), extraterrestrials or humans arranged it to be so? Why should it be wrong to arrange for our own arrival?

Each of them is extremely unlikely, but nothing like as unlikely as the notion of random chance – the endless mega-string of beneficial good luck.

The idea of the Möbius principle is that it is a loop that twists back in time and returns forward again. Imagine a situation whereby an artefact (say a black monolith) was manufactured in the year 2010 and was taken back in time by four billion years in 2011, where it was buried in a location of long-term stability on the Moon. The artefact could be recovered from the Moon before it was manufactured and the atomic matter from which it was made would exist in two places at the same time, until it was transported back to the early Moon.

This seems impossible. But just about everything about quantum physics sounds implausible. Quantum physics tells us that everything from light to matter is made up of tiny, indivisible packets called quanta that do not work as we normally see the world. Niels Bohr, one of the pioneers of the subject said: 'Anyone who can contemplate quantum mechanics without getting dizzy hasn't properly understood it.'

One of the features of this branch of science is the recognition that particles (or wave functions) briefly exist in several different places simultaneously. The monolith that had two concurrent realities would be a quantum effect on a worldly scale instead of at a sub atomic level.

Once the 2010 artefact goes back in time, the duality will be resolved and the world will continue as normal. Equally, we could consider all of the time, from the building of the Moon through to the point of time travel, as a Möbius loop where we end up back where we started. Thereafter we break out of the loop and move forward in the normal way.

Time and again

We have speculated that chronobots were sent back to engineer the Moon and they must have returned again nearly a billion years later to seed the ploughed Earth with DNA, to begin the process of evolution that would result in the arrival of humans.

But it appears that there must have been other interventions at specific times in the past to bring about certain events.

We have always agreed with archaeologists who say that the existence of the Megalithic Yard is inconsistent with the technology otherwise known to have been present amongst the people of western Europe over five millennia ago. But we heartily disagreed with them when most chose to ignore Professor Thom's findings rather than attempt to reconcile them. Such people are obstacles to knowledge.

When we discovered that the Megalithic system extended to the Moon, our credulity was stretched to the limit but our curiosity carried us forward to try and make sense of that which looked impossible to reconcile. When we found that the metric system had been in place almost perfectly, four-and-a-half thousand years before a team of French scientists reinvented it, we were amazed. Then we discovered that metric units were perfect integers for the most crucial aspects of the Moon as well as the Earth.

We have noted that through ancient history different civilizations have recorded that people with super powers arrived from nowhere to teach humanity about the sciences. Then we noted how all the parallel developments that occurred around the world in unconnected locations happened at the same time.

We have to conclude that people will travel back to points in history, such as the era around 3100BC, when several civilizations, from South America to North Africa to Asia to Europe, were suddenly emerging and building similar structures. It seems probable that the Megalithic structures that have lunar alignments, and use the unit that describes the dimensions of the Moon, were deliberately designed and left to point the way forward.

We do not yet know whether the detailed message is indeed inside the protected sections of DNA, but wherever it is, the initial message was only recognized because of all those Megalithic structures extending their weathered and ancient stone fingers into the night sky.

The fact that the imperial pound and the pint are mathematically derivable from the Megalithic Yard was puzzling and when we found that the same Stone-Age unit describes metric spheres we were dumbfounded. How could such surprising consequences come about so accurately by chance alone?

It now seems that the past has been modelled by the future. A strange Möbian twist for reality.

Of course this all sounds so improbable that some people will refuse to believe it. They will reject the fact that everything we have put forward is real and testable and the elements of unavoidable speculation are scientifically sound. But many creationists will still shout that black is white and many so-called scientists will return to their deeply flawed paradigms as though they were real.

But when the message is actually found. What then?

CHAPTER FOURTEEN

The Möbius Mission

'My own suspicion is that the Universe is not only queerer than we suppose but queerer than we can suppose.'
Haldane's Law put forward by geneticist, J P S Haldane

We have come on a strange journey since we first realized the science that lay behind the stone circles at Stonehenge, Brodgar, Avebury and thousands of other Neolithic sites across the British Isles.

We believe that our determination not to draw conclusions too early has paid dividends. We have refused to ignore those pieces of the puzzle that seemed outlandish or even downright impossible, and have retained our tolerance for unexpected results.

They say all progress is dependent on the unreasonable person. Alexander Thom was certainly an unreasonable person, or else he would have capitulated in the face of the wave of rejection he received from the majority of professional archaeologists. How irritating of the man, his opponents thought, to repeatedly insist that his reams of data, gathered over scores of years, show that the Megalithic builders worked to an incredible degree of engineering accuracy and employed precise standardized units of length.

Any reasonable person, and certainly any academic who wanted future employment within a specific discipline, would have buried the data that showed that the Megalithic Yard and its accompanying geometry were integer to the Moon and Sun as well as the Earth. It sounds ridiculous, and those who are members of the 'club' will consider anyone who speaks of it equally ridiculous.

Yet the Moon unquestionably does conform to Megalithic geometry and now, we believe, we are beginning to understand why.

The SETI institute is still sweeping the skies looking for incoming electromagnetic radiation that deviates even a tiny amount from the anticipated background noise – something that could conceivably be an indication of intelligence elsewhere in the cosmos. But the message they seek is already with us because, by the standards of what SETI considers might constitute a message, surely the material we have described here must be the world's first prime contact with a consciousness that existed 4.6 billion years ago.

The assumption of SETI and its operatives is that another intelligence will make contact from far away and so the searchers are focused on staring into the far depths of space for a message. But if any entity were that smart why would it have to make a long-distance call?

Four hundred years ago our solar system was the great mystery – but our own immediate environs hold less fascination in the light of incredible devices such as the Hubble telescope, which can reach far into space and into the past. We have ticked the box that is the solar system and astronomers are more interested in distant quasars and nebula. Is it this new leading edge of attention that has previously blinded us to the obvious in our own backyard?

The message we have received has told us about the way that the Moon was constructed to give life to the Earth, and there are tantalizing hints that this design may extend to the rest of the solar system, and possibly even beyond. Why is Jupiter in such an untypical orbit that just happens to be a cosmic umbrella for Earth? Why does Venus provide such a perfect clock and calendar when viewed from Earth?

Modern scientific culture has evolved from its roots in the ancient world and has become a complex web of many highly specialized disciplines. Gone are the days when one man, such as the seventeenth-century Robert Hooke, could be a groundbreaking inventor, microscopist, physicist, surveyor, astronomer, biologist and even artist. Today the sheer enormity of available information has led to highly defined specialisms, and academics are expected to keep to their field – despite the truism that science has no experts. No one, for instance, doubted Alexander Thom's abilities as a professor of engineering but he was not welcome in the world of archaeology.

The gains from modern science are beyond counting. But the loss, arguably, is the synthesis of information generated by the many gentleman scholars that once existed, before becoming extinct somewhere around the late nineteenth century. So few scholars now have a chance to view the bigger picture – to seek out patterns that might unexpectedly exist when apparently unrelated data is brought together. It has to be remembered that the difference between a major breakthrough and nothing at all can be just the angle of view rather than anything else.

Occasionally, two or more disciplines are brought together to form a new speciality in science. One of these turned out to be the subject that was directly invented by Alexander Thom, namely archaeoastronomy – a field of study involved with the use of astronomy by ancient cultures. Our previous book, *Civilization One,* demonstrated the geometry that lies behind Thom's proposed Megalithic Yard. We unambiguously showed how it is directly related to other measurement systems (linear, volumetric and weight) and put forward a testable theory of how it was reproduced using

Venus and a pendulum. We therefore decided to send a copy of our book to a man who we believe is the world's only professor of archaeoastronomy. He received a précis and the completed book but we received no response whatsoever.

We knew that the information we had put forward was not incorrect because people suitably qualified in astronomy and mathematics had carefully checked it. So why no response? Perhaps the approach was so counter to the worldview of this particular expert he could not understand it. Or maybe he just did not like the implications of our conclusions.

We also attempted to get a copy of the book to a world-class physicist. When he was told of the subject matter he responded almost angrily by saying that it was well known that Thom's work had been discredited decades ago and only weirdos clung onto the romantic hope that Stone-Age man possessed a rational and unchanging unit of length.

In actual fact he was repeating a mantra that has sprung up in academic circles but is no more than an urban myth, because no one has proven Thom to be wrong. We responded by pointing out that we had done our research very carefully and that whilst there are certainly people who have argued against Thom's conclusions, they have not proven his conclusions to be wrong – nor is it possible for anyone to prove a negative.

The academic then responded politely and accepted what we said, although he explained that he did not have time to read our book due to personal problems.

We therefore expect to have a fight on our hands when it comes to getting leading academics to review the findings contained in this book. But fight we will.

Finally, it is probably helpful to apply to our whole hypothesis a test created by the late, great astronomer Carl Sagan, that he called a 'Baloney Detection Kit'. Sagen suggested a set of tools shown below for testing claims and detecting fallacious or fraudulent arguments. We have put our responses beneath in italics.

Wherever possible there must be independent confirmation of the facts.

All of the key elements that we consider constitute a message are checkable using data published by leading authorities.

Encourage substantive debate on the evidence by knowledgeable proponents of all points of view.

Yes please. We have tried and will continue to do so.

Arguments from authority carry little weight (in science there are no authorities).

At least it should be a level playing field.

Spin more than one hypothesis – don't simply run with the first idea that caught your fancy.

We have had to dismiss only one possible hypothesis: coincidence, and have investigated every other avenue we can think of.

Try not to get overly attached to a hypothesis just because it's yours. Quantify, wherever possible.

We struggled to accept our own results initially and we remain entirely open to any other interpretation that might be brought forward.

If there is a chain of argument every link in the chain must work.

Unlike the existing main theories of the Moon's origin and the origin of DNA, we believe that we have a very strong chain with no weak link.

Occam's razor – if there are two hypotheses that explain the data equally well, choose the simpler.

Absolutely. But the simplest is also the weirdest, although it is also the most scientifically robust.

Ask whether the hypothesis can, at least in principle, be falsified (shown to be false by some unambiguous test). In other words, is it testable? Can others duplicate the experiment and get the same result?

The number sequences we have found are checkable by anyone with a book on basic astronomy and a calculator. The question is just how far people are prepared to go in claiming coincidence.

We believe we have made a case that deserves to be heard and investigated. It is hard to imagine how even the most sceptical, unimaginative academic could deny the possibility that we are looking at a message here. Not to investigate and discuss it further would be anti-scientific and, we believe, very foolish.

Everyone who has bothered to think about it is agreed; any message from the distant past either has to be very big or very small. We believe it is both.

We have good reason to think that Professor Paul Davies and others who suspect that there could be a communication addressed to us in apparently empty sections of DNA are correct. If the next layer of the message is, as seems likely, contained in the cells of our own bodies – it must be sought out!

Our suggestion that the group of a million unused base pairs, found by geneticist Edward Rubin and his team, might be a viewable message if laid out on a format of 3663x273 has to be tested. It might be the answer and, if so, humankind is on the verge of a new age – an age of maturity.

But if the message is not detected in that way, the peculiarities of the Moon remain and we need all serious scientists to work together to solve this riddle – which must be almost in our grasp.

We call on the world to assemble a team of leading scientists from all of the disciplines that could possibly be involved in deciphering the Moon's message and, if our third scenario is correct, constructing the CTC – the time transport system. And we may need observers from the leading religions.

We suggest that this be called 'The Möbius Mission' – a project to begin all projects!

Albert Einstein was an incredibly wise man as well as a scientific genius. Amongst his many quotable observations he once said: 'Imagination is more important that knowledge.'

How true. We therefore need people with depth of vision as well as knowledge and practical ability. So, it is our intention to first approach scientists such as Paul Davies, David Deutsch and Ronald Mallett. We feel sure that their curiosity will help to change the world.

APPENDIX ONE

Using the Megalithic Pendulum

About Pendulums

A pendulum is one of the simplest devices imaginable. In its most basic form it is nothing more than a weight suspended on a piece of twine or sinew. Since the pendulum has another function, as a plumb line, it may well be one of the first devices used by humanity. If allowed to hang, the weight of a pendulum will pull its string into a perfectly vertical position. Certainly the Megalithic people could never have constructed any of the major sites to be found all over Britain, Ireland and Brittany without the use of this device. It is therefore reasonable to suggest that if they possessed a plumb line, they also possessed a pendulum.

Although the device had been around for a long time it was the sixteenth-century genius Galileo who seems to have been the first person to look seriously at the attributes of pendulums (or at least the first of whom we have a record.) He is reported to have been bored in Church one day when his attention was caught by a large incense burner suspended from above by a chain or a rope, gently swinging back and forth and forming a natural pendulum. Galileo realized that the swings of the pendulum were equal in terms of time and he counted them against the beat of his own pulse.

Only two factors are of importance in the case of a simple pendulum. These are the length of the string and the gravitation of the Earth, which constantly exerts a pressure that will eventually bring the pendulum back to a vertical and resting position. The height of the swing of a pendulum is, to all intents and purposes, irrelevant because its time period from one extremity to the other will always be the same. In other words if the pendulum is excited more vigorously it will swing higher but its time period will remain the same.

It was a recognition of this constant nature of a pendulum that made it useful in the creation of clocks. In modern timepieces the pendulum has been superseded, but for many centuries it ensured the smooth running of clocks all over the world. It can still be found in quality clocks. Clock pendulums were eventually fitted with devices to prevent them from swinging too high, and others to regulate the nature of their arc of swing, but they are still, essentially, only animated plumb lines.

The Megalithic Yard

The Megalithic Yard was discovered by Alexander Thom as part of the composition of Megalithic sites from the northernmost part of Scotland, right down to Brittany in the South. The main problem with its use, and the reason archaeologists still doubt its veracity, lies in the fact that it remained absolutely accurate across thousands of square miles and many centuries. This would appear to be impossible in the case of a culture that was, at least in its early stages, devoid of metals to make a reliable 'standard' against which others could be set. Alexander Thom himself could think of no reliable way of passing on the Megalithic Yard without some variation being inevitable across time.

We eventually reasoned that it would be possible to turn 'time' into 'distance' by way of the turning Earth. The speed of the Earth on its axis is as accurate a yard stick for the passing of time as anyone could reasonably require for most purposes. Of course we can't see the Earth turning but we can see its effects. The Sun, Moon and stars appear to rise from below the horizon in the east, to pass over our heads and then set in the west. In fact, although the Moon and planets do have independent movement, the Sun and the stars are not really moving at all (actually they are moving slightly but we need not concern ourselves with this for our present purposes).

The apparent motion of the stars is caused by the Earth turning on its axis and it is this fact that offers us an accurate clock which, with a little ingenuity, we can turn into a replicable linear unit of measurement. In the case of the Megalithic Yard we eventually discovered that the pendulum upon which it is based was set not by viewing any star but the planet Venus. Venus is, like the Earth, orbiting the Sun. As a result, when seen from the Earth, it has a complex series of movements against the backdrop of the stars. Sometimes Venus rises before the Sun, at which times it is called a morning star, and at other times it rises after the Sun and is then known as an evening star. This is purely a line-of-sight situation, caused by the fact that both Venus and the Earth are orbiting the Sun. When Venus crosses the face of the Sun to become an evening star, it is moving 'against' the direction followed by the backdrop of stars. It is within this observable fact that setting the Megalithic pendulum becomes possible.

In order to create the Megalithic Yard one has to follow the simple rules below:

Venus must be observable as an evening star, setting after the Sun and during that period at which it is moving at its fastest counter to the backdrop of stars.

The sky is divided into 366 parts. This can be achieved by trial and error, as explained in *Uriel's Machine*[44] and also in *Civilization One*[45] but is also achievable through a neat little mathematical trick demonstrated below.

1 Stand in an unobstructed position on a wide open piece of ground with a good view of the western horizon.

2 Place a stick in the ground (stick A) and stand facing west with one of your heels touching the stick.

3 Now take 233 steps, heel to toe, towards the west. Upon completing the 233 steps, place a second stick in the ground (stick B) in front of your toe.

4 Turn to the north and place your heel against stick B. Now take four heel-to-toe steps to the north and then place a third stick (stick C) in the ground in front of your toe.

5 The distance between sticks B and C, when viewed from A will now be 1/366th of the horizon.

It is now necessary to make a braced wooden frame of the type shown in figures 14 and 15, which is as wide as the gap between B and C. This must be set on poles in such a way that it gains significant height and can be altered in its angle. The purpose of this exercise is so that the angle of the braced frame can be identical to that of the planet Venus as it falls towards its setting position.

Standing at A it is now necessary to observe Venus passing through the gap in the braced frame whilst swinging a pendulum and noting the number of swings achieved as Venus passes through the gap. A pendulum that swings 366 times during this occurrence must be 1/2 of a Megalithic Yard in length (41.48cm). The cord of this length represents the full Megalithic Yard of 82.966cm in length.

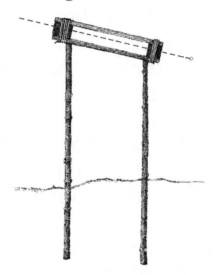

Figure 14

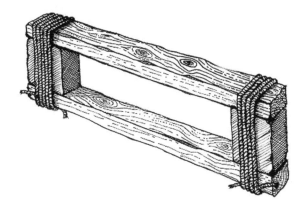

Figure 15

In this way the Megalithic Yard can be reproduced on any site where observation of Venus, when at the right part of its cycle, can be achieved. For the use of the braced frame we are grateful to the considerations of Professor Archie Roy, Emeritus Professor of Astronomy at Glasgow University.

Although pendulums differ slightly with latitude and altitude, because gravity also alters slightly, we have shown that the Megalithic Yard achieved using this method will remain within the tolerances discovered by Alexander Thom from Orkney in the north to Brittany in the south, in other words across the whole area containing monuments surveyed by Alexander Thom.

APPENDIX TWO

Using the Sumerian Pendulum

The method used by the Sumerians to set their own basic unit of length, the double kush, followed the same general rules as those employed by the Megalithic peoples of the far west of Europe. The only difference lay in the numbers used.

Sumerians relied on a 360° geometry, of the type we still use today. Because of this their starting point was to divide the horizon into 360 equal units. The mathematical trick used to short-circuit this procedure that was itemised in Appendix One does not apply in this case. It is possible that the Sumerians devised their own method of making the initial procedure quicker, but in any case theirs was a metal-using culture and one that would therefore not have needed to repeat the procedure of defining the linear unit all that often. They could have created a fairly accurate standard rod. Establishing the necessary 1/360th of the horizon by trial and error would have taken time, but it is quite possible to achieve with a high degree of accuracy.

The procedures used in the preceding Appendix are now followed. The braced frame would be equal to a gap of 1/360th of the horizon but Venus would be tracked in exactly the same way. The desired number of swings in this case is 240, which is the same as 240 seconds, a period of time known to the Sumerians as a 'gesh'. A pendulum that swung 240 times during the passage of Venus through the braced frame would be 99.88cm in length, a linear length that conforms to that discovered on the statues of Gudea from Lagesh in Iraq. This unit of length was known to the Sumerians as the double kush.

It has to be noted that the pendulum in question is not strictly speaking a seconds pendulum of the sort that was popularly used from the seventeenth to the nineteenth century. Because the object being tracked is Venus, which is moving independently against the backdrop of stars, the time taken for each beat of the pendulum is slightly longer than a second (1.002 seconds). This stands as part of the proof that the Sumerians did use this system to define their linear unit. They fully understood that there were 43,200 seconds in a day (to us there are twice this number because we use a twenty-four-hour day instead of the Sumerian twelve-hour day) but there is no absolutely reliable way of defining the true second of time by observing the sky and swinging a pendulum. This could only be achieved by tracking the average movement of the Sun in the same way Venus is used in this exercise. However, because of the Earth's own orbital characteristics, the Sun does appear to move at a constant speed across the sky. There are only a few days each year on which the experiment using the Sun would work perfectly and the Sumerians could not have known which days would have been appropriate. In addition, the Sun is very much more difficult, and potentially dangerous, to track in this way.

Similarly, if they had used a star instead of the planet Venus, the pendulum would still not have been a true seconds pendulum. The reason for this is that the sidereal day (a day that is measured by a star passing from one point in the sky back to that point again) is shorter than a solar day (a day that is measured by the Sun passing from one position in the sky back to that point again). A seconds pendulum created by tracking a star would actually give a time reading of 0.997 seconds and lead to a pendulum length close to 99.3cm.

We remain convinced that both the Megalithic culture and that of the Sumerians were simply following instructions that had been given to them by another agency. In the case of the Sumerians the use of Venus for setting their pendulum, and therefore their chosen unit of length, resulted in a series of measurements that were truly integrated with the Earth. As we have shown, the Sumerian double mana unit of mass divides into the mass of the Earth 6,000,000,000,000,000,000,000,000 times, which would not have been the case with a shorter pendulum and therefore a lighter unit of mass. True, the achieved second of time was slightly at odds with the genuine second of time, but the Sumerians, lacking accurate clocks, could not have been aware of this fact. In fact the discrepancy is so small it couldn't have been measured until the last century or so.

APPENDIX THREE

The Message in Detail

The message that we have detected is present in recurring number sequences that are, strangely, often round numbers. We started to realize that something highly unusual was happening when we discovered that the Megalithic system of geometry worked on the Moon and the Sun as well as the Earth.

Looking into issues concerning the Moon we were immediately reminded of the strange coincidence that the Moon and the Sun appear to be the same size in Earth's skies, leading to the phenomenon we call a total eclipse. Still stranger is the fact that the relation is so numerically neat with the Moon being 400 times smaller than the Sun and 400 times closer to the Earth at the point of a total eclipse. On its own this could be a bizarre coincidence, but because of what follows we believe that it is the 'headline' to a message built into the Moon 4.6 billion years ago.

The Megalithic system

The Megalithic system of geometry is based on 366° to a circle, sixty minutes to a degree and six seconds to a minute. This sequence produces a second of arc on the Earth's polar circumference that is 366 Megalithic Yards long, the linear measure of the Megalithic builders as identified by Alexander Thom.

As a cross-reference we had also discovered that the 4,000-yearold Minoan Foot is precisely equal to a 1,000th part of a Megalithic second of arc.

We applied the principles of Megalithic geometry to all of the planets and moons in the solar system and found that it only produced round integer results for the Sun and the Moon.

The Sun is very close to being a true sphere, certainly much more so than the Earth. NASA quote the mean volumetric circumference as being 4,373,096km, which we converted into Megalithic Yards and applied the 366 geometry.

Sun's circumference	=	5,270,913,968 MY
One degree	=	14,401,404 MY
One minute	=	240,023 MY
One second	=	40,003.8 MY

The fit is 99.99per cent accurate to 40,000 and given that this is based on a best estimate of the mean circumference it has to be considered bang on.

Like the Sun, the Moon is quite close to being a sphere. NASA gives the mean volumetric circumference of 10,914.5km, which produces the following result:

Moon's circumference	=	13155300 MY
One degree	=	35943 MY
One minute	=	599 MY
One second	=	99.83 MY

If we use the equatorial radius the result is 99.9 MY per second of lunar arc. Either way, this is as close to 100 MY as makes no difference, given the irregular surface of the Moon and the small variation in Thom's definition of the Megalithic Yard of +/- 0.061cm.

It could have been possible for people many thousands of years ago to create a system of geometry that produces round integers for two celestial objects such as the Earth and the Sun, but it would seem impossible to achieve such a feat for three bodies. It therefore appears that the Moon was designed using units derived from the physical dimensions of the Sun and the Earth.

The Earth–Moon relationship

The duration of the Moon's orbit (sidereal – fixed star to fixed star) is 27.322 Earth days (27.396 rotations of the Earth). This number is extraordinarily close to the size relationship of the Moon to the Earth, being 27.31 per cent of the Earth's size.

The Earth currently turns on its axis 366.259 times for each orbit around the Sun. This number is extraordinarily close to the size relationship of the Earth to the Moon, being 366.175per cent larger than the Moon.

There is no reason why these numbers should repeat in this way:

	Earth turns per orbit	per cent size of polar circumference
Earth	366.259	27.31
Moon	27.396	366.175

It is also a consequence of the above that the Moon makes 366 orbits of the Earth in 10,000 Earth days.

The size of the Sun, Earth and Moon have been fixed for billions of years so their size ratios have not changed. But the orbital characteristics of the Earth and the Moon have changed constantly.

When the Moon was much closer to the Earth than it is now, its orbit was much shorter and the Earth day was also shorter, leading to perhaps as much as 600 days to the Earth year. The Earth's own orbit around the Sun remains essentially unchanged. It is only the time it takes to spin on its own axis that alters.

The close number association between the size ratios of the Sun, Moon and Earth, and the orbital characteristics of the Moon, together with the present length of the Earth day, are only applicable to the time that humans have been fully formed. These relationships were not present in the distant past and they will disappear in the distant future. The number sequences which alerted us to the 'message' are clearly meant for the present period.

The Metric System

Orbital characteristics and size relationships are physical factors and any correlations are real – no matter what units of measurement are employed. No one knows the origin of the Megalithic system but the origin of the metric system is fully documented. Whilst it did have a near identical precursor in the Sumerian system of more than 4,000 years earlier, the metric system is known to have been developed from measuring the polar circumference of the Earth alone.

It was designed so that there should be 40,000km in one Earth circumference. The equator is a little longer than the polar circumference but basically the Earth turns through this distance each day.

The Moon turns through an unimpressive sounding 10,920.8 kilometres every 27.3217 days. This converts to 400km per Earth day – to an accuracy greater than 99.9 per cent.

Again this is a factor that only exists in the human period of existence.

The number 400 is already central to human appreciation of the Moon because it is 400 times closer to us than the Sun, and it is 400 times smaller. The use of 400 kilometres per current Earth day could be a message that the architect of the Moon knew we would use kilometres and mean solar days.

Metric units apart, the Moon is turning at a rate that is almost exactly one per cent of the Earth's rotation. Or to reverse the factor, the Earth is turning 100 times as fast as the Moon. All curiously round values!

To add to the idea that this is a deliberate piece of metric design, the Moon is also travelling on its journey around the Earth at a steady rate of one kilometre per second! This speed varies a little as it travels but does not drop below 0.964km per second and does not exceed 1.076km per second.

And there is something else very special about the kilometre as regards the Moon. To understand it we need to realize that there are 109.2 Earth diameters across the Sun's diameter. There are also 109.2 Sun diameters between the Earth and the Sun at its furthest point of orbit.

The circumference of the Moon is 109.2x100 kilometres.

Is that not odd in the extreme?

One way of looking at the association between these ratios and numbers can be seen in the diagram in figure 16.

There are many factors here that should bear no relationship with each other at all. Taken in isolation, any one of these strange associations might be considered to be a coincidence but there comes a time, however, when coincidences become so frequent that something else must be at work.

366
The number of rotations in an Earth year

366
The number of Megalithic Yards in 1 Mg second of arc of the Earth

366%
The percentage size Moon to Earth

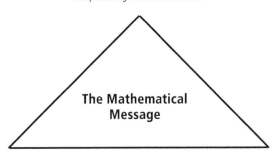

The Mathematical Message

400
The ratio of the size of the Moon
to that of the Sun

109.28
The ratio of the size of the Earth
to that of the Sun

1/400th
The number of times the Moon is
closer to the Earth than the Sun

109.25
The number of Earth diameters
across the diameter of the Sun

40,000
The number of Megalithic Yards in
1 Mg second of arc of the Sun

109.26
The number of solar diameters
across the Earth's orbit at aphelion

40,000
The number of kilometres the Earth
turns on its axis in a day

27.322
The sidereal days in 1 lunar orbit
27.322 X 4 = 109.2

400
The number of kilometres the Moon
turns on its axis in a day

27.322%
The percentage size Earth to Moon

10,000
The number of days in 366
lunar orbits

10,920.8
The size of the Moon in kilometres

100
The number of Megalithic Yards in 1 Mg
second of arc of the Moon

400
The number of times the Earth
rotates faster than the Moon

Figure 16

APPENDIX FOUR

The Mechanics of Eclipses

The awesome sight of a black shadow gradually crossing the face of the Moon still captivates most people, even though we now live in an age when we not only know what causes the phenomenon but can predict exactly when it is likely to happen. Early cultures did not know either and must have seriously thought, for a few minutes at least, that the world was coming to an end.

Back in the 1960s the astronomer Gerald Hawkins suggested that at least one of the functions of the structure at Stonehenge, Salisbury Plain, England, was to predict the occurrence of eclipses. Hawkins had carefully studied the ancient monument, parts of which date back five thousand years, and subjected his data to a massive number-crunching computer. He came to the conclusion that the Aubrey Holes, a series of fifty-six chalk-filled pits around the standing stones at Stonehenge, represented a sophisticated device for predicting both solar and lunar eclipses.[46]

Clay tablets discovered in what is now Iraq and dating back to the Sumerian period, which commenced around 3000BC, indicate that people in the region were doing their best to predict eclipses. And there isn't any doubt at all that the Babylonians who followed the Sumerians were competent at accurately working out when the face of the Sun or Moon would darken.

The ancient Chinese, Indians, Egyptians, American cultures and many other societies worked hard to develop an understanding of rudimentary astronomy for the purpose of eclipse prediction. This single effort certainly caused humanity to significantly improve its naked-eye astronomy and its understanding of mathematics. There are good reasons why this should be the case and at the base of most of them is power. Any would-be ruler, secular or religious, who could predict when an eclipse was likely to take place was in a very strong position to manipulate the situation to his or her own ends.

To the average lay person, eclipses seem to be totally haphazard but this is not the case. However, such is the complicated nature of the interplay of the Earth and the Sun that understanding eclipse patterns is far from easy. Once the pattern is cracked, its secrets could be passed from one ruler to another and a whole society could be alerted to a possible eclipse. The prediction itself would have seemed to most people to be the most sophisticated sort of magic and when the king or holy man drove away the dark dragon that was trying to swallow the Sun or causing the Moon to bleed, his power would be ensured for a considerable period ahead.

What the ancients gradually discovered was that there were very definite patterns to the occurrence of all eclipses and that they were governed overall by a specific period of time that is known as the 'saros cycle'. The word saros was first used by the astronomer Edmond Halley (1656–1742) and is supposed to have been derived from a Babylonian word. The saros cycle is 6,585.3 days in length (18 years, 11 days, 8 hours). It represents the coming together of three distinct patterns. The first of these is the Synodic Month (new Moon to new Moon,) the second is the Draconian Month (node to node [see below for information on Moon's nodes]) and the third is the Anomalistic Month (perigee to perigee [see below for information on Moon's perigee]).

To within about two hours, 223 synodic months, 242 draconian months and 239 anomalistic months come out to the same period of time and it is at this point that any eclipse will repeat itself. The reason for this is that the solar system runs pretty much like a gearbox and, as with a gearbox, any pattern created now will sooner or later be repeated.

Although the saros cycle is very accurate, there are many such cycles running at the same time. All that can be deduced from the saros cycle is that if an eclipse occurs today, it will occur again in 6,585.3 days and will have a quite similar geometry. The system does fall down somewhat in that it splits a day and so future eclipses in any given cycle may not be fully visible from the same part of the globe. Each saros cycle runs for around 1,200 years (around sixty-six repeat eclipses) until it expends itself. If the saros cycle commences near the South Pole it will extend itself gradually further north with each eclipse until it finally disappears at the North Pole. The same is true in reverse.

It would appear that the Babylonians understood the saros cycle, as did the Ancient Greeks. So, according to Gerald Hawkins, did the builders of Stonehenge. Something akin to the saros cycle would have been useful to ancient peoples because if the next eclipse in any given series was less impressive than the last, it would still have been predicted, and it was just as likely to be more impressive as less so. (Better by far to turn the tribes out for a less-than-spectacular display than to miss what could be a super show!)

Our own previous research demonstrates that following the saros cycle was actually very easy for the Megalithic people, who were the builders of Stonehenge and thousands of other such monuments. The ritual year of the Megalithic cultures was 366 days in length. This meant that the saros cycle to them was just two days short of eighteen years in length. The two days didn't really matter because solar eclipses can only occur at the new Moon and lunar eclipses at the full Moon. In other words, just a couple of days short of eighteen years after a particular eclipse, the next full or new Moon would be certain to bring another.

Even today we don't take solar eclipses for granted. A major eclipse, such as the one that was visible in northern Europe on August 11th 1999 is treated as a time of celebration and is now revered for its sheer beauty, rather than being feared as was surely the case even not so long ago. The face of the Sun gradually begins to blacken as the Moon passes between it and the Earth. If it is a full eclipse the Sun's disc will be covered at what is known as totality. At totality, all that is perceptible is the faint glow from the corona of the Sun. Soon after, the shadow begins to move away and a spectacular shaft of light breaks out, forming what is known as the diamond ring effect. The phenomenon is just as impressive now as it must have looked from Babylon or Stonehenge.

It might surprise readers to learn that no matter where our astronauts or cosmonauts travel in the future within our solar system, they will never stand on the surface of any other planet and watch a total eclipse. They are simply not possible anywhere else and only occur as a legacy of a series of breathtaking, apparent coincidences. The fit of the Moon's disc across the face of the Sun during a total eclipse is not 'near' – it is 'exact' – and this fact should be the greatest sense of wonder to anyone viewing such an event because it is very unlikely. No other planet has a moon anywhere near big enough or orbiting at the right distance to fully, but not too fully, eclipse the Sun.

There are two basic sorts of eclipse, and then subcategories within the two types. The most impressive form of eclipse is known as a solar eclipse. The drawing below shows what is actually happening when a solar eclipse takes place.

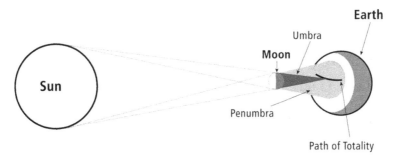

Figure 17. When the Moon stands directly between the Earth and the Sun, a total eclipse is possible but totality only occurs across a relatively small area of the Earth's surface and follows a curve known as the Path of Totality

In this example, which is a 'total eclipse', to a proportion of those people living along the path of totality, the disc of the Sun will be blotted out completely. Whilst totality is achieved, all that can be seen is the sun's corona (the halo of bright matter that is constantly being thrown off by the Sun). The larger shadow is called the penumbra and people beneath this will see a partial eclipse. There is another form of solar eclipse that can never be total and this is known as an annular eclipse. The Moon is 1/400th part the size of the Sun and it stands at 1/400th the distance between the Earth and the Sun, but not always exactly.

The Moon's orbit around the Earth is not circular but elliptical. This means that sometimes the Moon is slightly closer to the Earth than it is at other times. If a solar eclipse takes place when the Moon is furthest from the Earth, the Moon's disc looks smaller and can never totally blot out the Sun. Total eclipses of the Sun therefore happen when the Moon is on the part of its orbit that brings it closest to the Earth. When the Moon is closest to Earth it is said to be at *perigee* and when it is furthest away it is at *apogee.*

Solar eclipses can only take place when the Moon stands between the Earth and the Sun and this is the short period on each lunar cycle known as 'new Moon'. (The time of the lunar month when no part of the Moon is visible from Earth.)

It might be thought that because there is a new Moon each month, there should therefore be a solar eclipse each month but this is not the case. The orbit of the Moon around the Earth does not follow the same angle as the orbit of the Earth around the Sun. If it did, every new Moon would indeed bring a solar eclipse. Rather it is tilted to the Earth's orbit (known as the ecliptic) by five degrees. Only when new Moon occurs at a point when the orbit of the Moon around the Earth crosses that of the Earth around the Sun, can a solar eclipse take place. These points north and south of the ecliptic are called the Moon's nodes. This happens 'at least' twice each year and can produce a solar eclipse observable from somewhere on the Earth.

The second type of eclipse is not quite so impressive as a solar eclipse but it would have been fascinating to our ancient ancestors all the same. It is more common than a solar eclipse and is known as a lunar eclipse. A lunar eclipse takes place when the shadow of the Earth comes between the Sun and the Moon. A lunar eclipse can only take place at the exact opposite time to a solar eclipse, at the time of the full Moon when the entire disc of the Moon is visible from Earth.

During a lunar eclipse the face of the Moon does not disappear altogether. Rather it is darkened and, under some circumstances, it appears to turn a deep red. Such lunar eclipses were seen by many ancient cultures as terrible harbingers of disaster and were probably feared as much as solar eclipses.

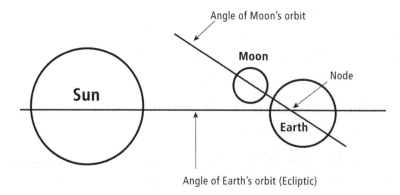

Figure 18. The path taken by the Earth around the Sun is not the same as that taken by the Moon around the Earth. There is a 5° difference. Because of this, total eclipses can only happen when new Moons fall on what is known as the node – that point at which the two orbits cross.

Figure 19. A lunar eclipse takes place when the Earth's shadow crosses the face of the Moon at the time of full Moon. Once again the fact that the plane of the Earth's orbit around the Sun and that of the Moon around the Earth are not the same prevents every full Moon from being eclipsed.

Until we did some in-depth research we never realized just how unlikely or extraordinary a total eclipse actually was. It's all a matter of 'line of sight' as the diagrams below should make clear. Isaac Asimov, the famed science-fiction guru, described this perfect visual alignment as being: 'The most unlikely coincidence imaginable'.

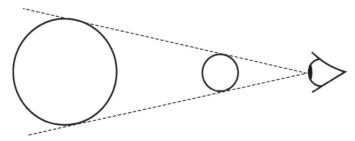

Figure 20. In this example the eye on the right looks past a small sphere to a much larger sphere. The size of the spheres and the distance between them is such that because of the perspective to the viewer the small sphere will exactly cover the large sphere.

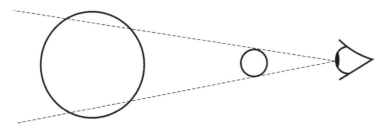

Figure 21. Now the small sphere is even smaller, but is at the same distance from the eye of the observer. Under these circumstances the eye will also see part of the larger sphere. Finally, if we keep the spheres the same as in the last example, but move the smaller one nearer to the eye of the observer we once again create a situation in which the small sphere appears to exactly cover the large one.

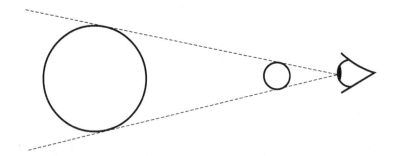

Figure 22. The sphere of the Sun is almost exactly 400 times larger than that of the Moon. This in itself might be considered nothing more than a strange but meaningless coincidence but we must stretch coincidence almost to breaking point when we realize that when the Moon is as close to the Earth as its orbit will bring it, it stands at 1/400th the distance between the Earth and the Sun. Under these circumstances when it stands precisely between the observer and the Sun, the Moon 'must' exactly cover the disc of the Sun – it is a simple matter of perspective.

In the case of total eclipses we really are living in a tiny snapshot of the history of the Earth and the Moon. The Moon was very much closer to Earth at the beginning of the relationship, and by the time the two reach a situation of perfect stasis the Moon will be 1.6 times further away from the Earth than it presently is. If we estimate the Moon to be 4 billion years in age and then accept the most common assessment that it will reach its furthest position from the Earth in 15 billion years (excluding the fact that the Sun will most certainly have gobbled up both Earth and Moon by then) the sum total of the Moon's journey from closest to furthest from the Earth is 19 billion years. The Moon is a finite size, as to all intents and purposes is the Sun. There can only be a very short window of opportunity during which the disc of the Moon can cover that of the Sun, as seen from Earth, in the truly perfect way that it does right now. That it has done so just at the time we have evolved into a sophisticated enough species to recognize and study the fact seems almost incredible. It doesn't matter how much experts say 'It's just one of those things', it is still an example of one of the most unlikely coincidences imaginable.

It stands to reason that if the Moon were any larger or smaller than it is, total solar eclipses would not be possible at this time. A smaller Moon would have brought such phenomena in the very distant past, when the Moon was much closer to the Earth than it is now.

This brings us to a discussion of the dimensions of the Moon. In comparison with the size of its host planet the Moon is huge. Its circumference would only fit into that of the Earth 3.66 times. Another of the terrestrial type planets, Mars, has two moons, but they are tiny when set against Earth's Moon, which in terms of size might reasonably be termed a planetoid. Even bearing in mind the vast size of the planetary super-giants, with their proliferation of moons, Earth's Moon is still the fifth largest in the whole solar system.

A close examination of Moon rock, brought back by both American astronauts and Soviet unmanned missions, shows that they are very similar to specific rocks on the Earth. Analysis of the rocks proves that they were created at the same distance from the Sun, so there is no longer any real doubt that the Earth and the Moon have a common origin. Yet there is something very strange about the Moon that isn't easy to explain. Although it is 1/3rd as big as the Earth it has only 1/81st of the Earth's mass.

Had the Moon been composed of a representative sample of 'all' Earth's rocks and still been the size it is, it would have been much more massive. Conversely, if the Moon had exactly the same composition as the Earth and had exerted the gravitational pull it presently does, it would have been very much smaller in size.

These facts are discussed in greater detail elsewhere, but it is the weird composition of the Moon, which is comprised of very light, Earth-type rocks, that means it can be large enough to create a total solar eclipse and yet not rip the surface of the Earth to pieces with its gravitational pull every time it passes overhead.

In a Universe filled with incredible wonders, and one so big that it might as well be infinite as far as we are concerned, we are certain to stumble across what looks to us like outrageous coincidences. Even conservative astronomers admit that total eclipses are very unlikely but still maintain such happenings must be a random chance event. We beg to differ!

APPENDIX FIVE

The World From a Barley Seed

The previous work we had undertaken for our book *Civilization One* had featured a number of ancient measuring systems. None of these surprised us more than that created by the Sumerians, a culture that originated in what is presently known as Iraq at about the same time as the Megalithic culture was flourishing in Britain and France. Our ongoing work for the present book made us look again at some aspects of the Sumerian measuring system. It could be that yet another part of the message left to us, indicating a deliberate intervention into the origin and progress of humanity, is encapsulated within the methods the Sumerians used to measure their world.

Out of a plethora of different linear lengths, weights and measures, it was possible for us to reconstruct the entire Sumerian system as we are sure it was originally meant to be. We have demonstrated how the Sumerians used a pendulum and the planet Venus in order to establish the basic unit of linear length, which was known as the double kush. Existent statues of the Sumerian King Gudea demonstrate that the double kush was intended to measure

99.88cm. Units of volume and weight were derived from the double kush by creating a cube with sides of 1/10th of a double kush. The amount of pure water held by such a cube represented the *sila,* which was the Sumerian unit for measuring volume. The weight of this water was known as the *mana* or *mina* and was the Sumerian unit for measuring mass. How we untangled all of this from the Sumerian records is explained in detail in our book *Civilization One.*[47]

There seemed to be no doubt that the double kush had indeed been created by way of a pendulum and observations of the planet Venus but it was not the only way the Sumerian system could be recreated. Everything in the system also relied on the size, shape and weight of a humble barley seed.

To the Sumerians a barley seed was known as a *se.* Until our own investigations, many experts had believed that the use of the barley seed by the Sumerians for measuring purposes was probably an abstraction. It was generally considered that the Sumerians might originally have used such seeds (as was the case in ancient western Europe), but that as in the case of Europe the seeds ultimately came to be words representing sizes and weights that no longer related to barley seeds at all.

The Sumerians claimed that 360 barley seeds was the measure of the double kush, something that experts on the Sumerian culture actively deny or at best have totally ignored.

Our extensive investigations showed conclusively why this state of affairs had come about. Experts had undoubtedly assumed that if the Sumerians had used barley seeds as tiny units of length, they must have laid the seeds end to end. It is likely to be for this reason that it is now generally considered that the seeds themselves eventually lost all contact with units of measure, because when they are laid end to end they make no sense at all. However, we discovered that if the seeds were laid on their sides and front to back (as they may have been carefully strung on a necklace) they conformed absolutely to the Sumerian system.

We then went on to demonstrate, by practical experiment, that the Sumerians had also been quite correct in their estimation of the 'weight' of an average barley seed and we were staggered to discover that even modern barley seeds have almost exactly the same dimensions and weight as their Sumerian counterparts.

It has been possible to show that in terms of mass measurement the whole Sumerian system was irrevocably and seemingly quite deliberately tied to the overall mass of the Earth itself. We appreciate that this sounds absolutely absurd for such an early culture but when one sees the figures involved, there is no doubt about it.

According to Sumerian texts it was considered that there were 10,800 barley seeds to the unit of weight known as the 'mana'. The weight of water held in the double mana, assuming a double kush of 99.88cm and a cube with sides of 1/10th of this, would have been 996.4 grams.

The mass of the Earth is held to be 5.976×10^{24}kg. If we divide this by .9964 in order to establish how many double mana there are to the mass of the Earth, we arrive at 5.99759×10^{24} double mana. This number is so close to 6×10^{24} (99.99per cent) that this must surely have been the number intended. Since there are 10.800 barley seeds to the mana and therefore 21,600 to the double mana it is possible to show that the mass of the Earth is equal to that of 1.296×10^{29} barley seeds. This might not seem to be a particularly impressive number but it has some very important properties.

If we were to segment the Earth, as we might an orange, we would discover that each 1/360th segment of the Earth has a mass equal to 3.6×10^{26} barley seeds. A further split of sixty brings us to 6×10^{24} barley seeds and yet another split of sixty results in 1×10^{23} barley seeds, which can be expressed fully as 100,000,000,000,000,000,000,000.

The starting point of this exercise was an Earth mass of 6×10^{24} double mana for the mass of the Earth, which would have been highly significant in Sumerian terms since there's was a sexagesimal (sixty base) system.

For all the reasons explained in *Civilization One,* we cannot accept that this state of affairs is a coincidence. What we have with the Sumerian system is a fully integrated way of measuring length, volume, mass, area and time, using the same number bases in each case. The whole system can be constructed from a pendulum set by the movements of Venus across one degree of arc of the horizon or else from the bottom up with nothing more complicated than barley seeds.

The real question has to be whether or not the Sumerians themselves could have possibly known just how incredible their measuring systems were? We are left with the impression that the system would have been very useful in the marketplace and on the farm in order to ensure equity of measurement throughout Sumerian society but that it is highly unlikely that the Sumerian Priests could have known the dimensions of the Earth, let alone its mass. It is most likely that both concepts would have been absolutely alien to them.

This appears to be yet another example of direct and deliberate intervention into the development of humanity. In other words, as their own mythology demonstrates, someone 'taught' the Sumerians about weights and measures and told them the numbers to use. By so doing they supplied the Sumerians with one of the hallmarks of true civilization, namely an integrated and replicable measuring system. At the same time, the use of the barley seeds added to a significant series of messages about these events in prehistory that were intended for our consumption. Since it seems unlikely that a cereal grain as widespread and useful as barley could, by chance, behave in the way that it does in terms of its size and weight, it seems very likely to us that the crop was genetically engineered. It was used by the Sumerians for bread but also brewed into a beer that was drunk for many centuries across the whole of civilization.

Notes

[1]   Fox News: June 18th 2004
[2]   Heath, Robin: *Sun, Moon and Earth.* *Wooden* Books Ltd, 2001
[3]   Knight C and Lomas R: *Uriel's Machine.* Arrow, 2000
[4]   Shlain, L: *The Alphabet Versus the Goddess.* Arkana, 1998
[5]   Hawkins G: *Stonehenge Decoded.* Hippocrene Books, 1988
[6]   Graham, Joseph Walter: *The Palaces of Crete.* Princeton University Press, 1962.
[7]   A 'library angel' is a term used by researchers for the strange moments when the required information seems to search them out. It is, of course, just a function of chance because when you are involved with seeking out a great deal of material you are going to get very lucky from time to time.
[8]   For an excellent description of the Allais effect see: http://www.flyingkettle.com/allais/eclipses.htm
[9]   Hartmann, William K: *Origin of the Moon.* Lunar and Planetary Institute, Houston, 1986
[10]   Hartmann, W K and Davis, D R: *Icarus* Edition 24, Cornell University, 1975

[11]   Lissauer, J.J: 'It's not easy to make the moon', *Nature* Magazine 389, 1997

[12]   Ruzicka, A., Snyder, G.A. and Taylor, L.A.: 'Giant Impact and Fission Hypotheses for the origin of the moon: a critical review of some geochemical evidence', *International Geology Review* 40, 1998

[13]   Knight, Christopher and Butler, Alan: *Civilization One. Watkins* Books London, 2004

[14]   Shklovskii, I S and Sagan C: *Intelligent Life in the Universe.* Emerson-Adams, 1998

[15]   Hood L L: The American Geophysical Union's Geophysical Research Letters. August 1st, 1999

[16]   Bhattacharjee C et al: 'Do animals bite more during a full moon? Retrospective observational analysis'. *British Medical Journal.* 2000 December 23; 321(7276): 1559–1561

[17]   Lieber A L: 'Human aggression and the lunar synodic cycle', *Journal of Clinical Psychiatry.* 1978 May; 39 (5):385-92

[18]   Laskar, J, Joutel, F & Robutel P: *Nature* 361, 1993

[19]   Comin N F: *What If the Moon Didn't Exist?* HarperCollins Publishers. 1993

[20]   Wegner, Alfred W, *The Origin of Continents and Oceans.* First Published

1915, publisher unknown. Published Dover Publications, Dover, 1966.

[21] N F Comins, *Voyages to Earth That Might Have Been,* HarperCollins New York 1993

[22] *Nature,* Genetics, DOI: 10.1038/ng1508

[23] Rennie. J: '15 Answers to Creationist Nonsense' *Scientific American.* July 2002

[24] Rose, C & Wright, G: 'Inscribed matter as an energy-efficient means of communication with an extraterrestrial civilization', *Nature* vol 431 p 47

[25] Jacobsen, Stein B: 'How Old Is Planet Earth?' *Science* Jun 6 2003: 1513–1514

[26] Haas J, Creamer W & Ruiz A: 'Dating the Late Archaic occupation of the Norte Chico region in Peru', *Nature* 432, 1020–1023

[27] Flew A: *How to Think Straight.* Prometheus Books. (2nd edition) 1998

[28] Lathe R: 'Fast Tidal Cycling and the Origin of Life'. *Icarus* 168(1) (2004) pp 18–22

[29] Yockey, Hubert P: *Information Theory and Molecular Biology.* Cambridge University Press, UK, 1992

[30] Dembski, William A: *The Design Inference: Eliminating Chance Through Small Probabilities,* Cambridge: Cambridge University Press. 1998

[31] Morowitz, H J: 'The Minimum Size of Cells', *Principles of Biomolecular*

Organization, eds. G.E.W. Wostenholme and M. O'Connor, London: J.A. Churchill. 1966

[32] Mora, Peter: 'Urge and Molecular Biology' *Nature, 1963*

[33] Bernal, J D: *The Origins of Prebiological Systems and Their Molecular Matrices,* 1965

[34] Hoyle, F: 'The Universe: Past and Present Reflections', *Engineering and Science,* November 1981.

[35] Hoyle, F: *Intelligent Universe.* M Joseph London, 1983

[36] Davies, P: 'The Ascent of Life', *New Scientist,* 11 December, 2004.

[37] Adams, D: *The Hitchhiker's Guide to the Galaxy.* Tor Books, London, 1979

[38] Hoyle, F: *The Black Cloud.* Buccaneer Books Inc (1992)

[39] Knight, C & Lomas R: *The Book of Hiram.* Arrow. 2004

[40] Chronobot: If we may be forgiven for inventing another word, this seems an appropriate term for a machine that labours across time.

[41] Deutsch, D & Lockwood, M: 'The Quantum Physics of Time Travel', *Scientific American,* March 1994

[42, 43] Deutsch, D & Lockwood, M: 'The Quantum Physics of Time Travel', *Scientific American,* March 1994

342

[44]   Davies, P: 'Do we have to spell it out?' *New Scientist,* 7 August 2004.

[45]   Knight and Lomas: *Uriel's Machine,* Transworld Books, London, 1999

[46]   Knight and Butler: *Civilization One, Watkins* Books, London, 2004

[47]   Hawkins, Gerald: *Stonehenge Decoded.* Hippocrene Books, London, 1988

[48]   Knight and Butler: *Civilization One. Watkins* Books, London, 2004

Lightning Source UK Ltd.
Milton Keynes UK
UKHW032115181218
334227UK00009B/261/P